RESPIRATION IN WATER AND AIR

Adaptations – Regulation – Evolution

RESPIRATION IN WATER AND AIR

Adaptations – Regulation – Evolution

Pierre Dejours

Laboratoire d'étude des régulations physiologiques
(associé à l'Université Louis Pasteur)
Centre national de la recherche scientifique
Strasbourg, France

1988
Elsevier
Amsterdam · New York · Oxford

ISBN 0-444-80926-0

Published by:
Elsevier Science Publishers B.V. (Biomedical Division)
P.O. Box 211
1000 AE Amsterdam
The Netherlands

Sole distributors for the USA and Canada:
Elsevier Science Publishing Company, Inc.
52 Vanderbilt Avenue
New York, NY 10017
USA

Library of Congress Cataloging-in-Publication Data

Dejours, Pierre.
Respiration in water air.

Bibliography: p.
Includes index.
1. Respiration-Regulation. 2. Physiology, Comparative. 3. Aquatic animals-Respiration. 4. Adaptation (Biology). 5. Evolution. I. Title.
QP123.D45 1988 591.1′21 87-27494
ISBN 0-444-80926-0

Printed in The Netherlands

To
Hermann Rahn

Contents

CHAPTER 5 – RESPONSES TO AMBIENT OXYGEN AND CARBON DIOXIDE VARIATIONS

CHAPTER 6 – RESPONSES TO VARIATIONS OF ENERGY NEEDS

CHAPTER 7 – INTERFERENCES BETWEEN RESPIRATION AND OTHER FUNCTIONS

CHAPTER 8 – HIERARCHY OF REGULATIONS

CHAPTER 9 – FROM RESPIRATION TO OTHER FUNCTIONS IN AQUATIC AND TERRESTRIAL ANIMALS

Introduction

The adaptation at every step of the CO_2- and O_2-transfer system to environmental changes and to variations of the energy intensity required by size, nutrition, temperature or exercise, has been extensively studied in many animals, but we know most about the external exchanges – branchial and pulmonary – which are more accessible to investigation. An enormous amount of morphological and functional data has been accumulated and presented in papers, reviews and books.

In the face of this bulk of knowledge two attitudes are often encountered. One consists of thinking or saying that everything is known, that any further study is superfluous; certainly this state of mind is not uncommon, but it is rarely acknowledged. On the other hand, it is common enough to read in the introduction to many articles that, in spite of many studies and reports, the subject matter remains obscure; it may even be affirmed that nothing is known. Actually, a lot has usually been done and understood; but it may give a flavor of originality and priority to proclaim that everything remains to be done. Neither of these attitudes is correct. Certainly, the amount of data is enormous; yet they are often insufficient; some need to be controlled or completed; some new data and, of course, new concepts open new roads to research. It is the aim of this book to try to present in order many physiological data which are rarely considered simultaneously.

Here, respiratory regulations are presented from the point of view of comparative physiology. The respiratory adaptations of animals to their environments and to their own changing needs will be described. Where possible, the regulatory mechanisms of these adaptations will be discussed, eventually in terms of control theory. My main thesis is that the oxygenation of the body seems to be the primary aim of the regulatory mechanisms and that the carbon dioxide clearance and the acid–base balance are subordinate requirements. In the last chapter I will try to show that for the physiologist there are some good reasons to divide the animal kingdom into aquatic and terrestrial animals; the respiratory characteristics suggest this division, and moreover the other contrasting properties of the two environments, water and air, which the animals dwell in, impose certain features of design and function.

Some of the views in this book are very probably correct. The present state of experimental study is nonetheless insufficient to guarantee the pertinence of certain statements I will insist on, which may quickly become obsolete or will have to be completely revised, or even abandoned. At least I hope they will be provocative.

Much of this book is accessible to any scholar, whatever his field of interest. But some parts may not be understood without some basic knowledge of physiology, particularly of comparative respiratory physiology. General references to books, reviews or important papers will be given, but for certain specific topics, a more complete bibliography is listed.

Symbols, abbreviations and units

The symbols follow the rules generally accepted in respiratory physiology since the Atlantic City Federation meeting of 1950 (Pappenheimer *et al.*, 1950); see also Dejours (1981) and Kellogg (1987).

Ideally all units used should be basic or derived units of the International System of Units (Bureau International des Poids et Mesures, 1985), namely:

quantity	*name*	*symbol*
length	meter	m
mass	kilogram	kg
time	second	s (but sec is better)
thermodynamic temperature	kelvin	K
amount of substance	mole	mol

The complete set of base units comprises also the ampere (A) and the candela (cd). Among the derived units, the pascal (Pa) is the SI unit of pressure. A liter should be named dm^3. We will not follow all the SI recommendations because what is more important than the strict observance of this system is the proper use of symbols and units with only one imperative in mind: there must be no ambiguity in the writing of symbols and use of units. Anyway the use of SI units cannot be general until the available tables of physical constants are rewritten accordingly.

I often use min (minute), L (for liter = dm^3), °C (for degree Celsius) which are mentioned officially as 'Units used with the International System' in the above-mentioned book. It is said (p. 75) that '... users of SI will also wish to employ with it certain units not part of it, but which are important and are widely used'. Furthermore I use almost exclusively the torr (symbol Torr) as the unit of pressure.

The table below lists the symbols used in this book, their meanings and examples of units. At the end of the book are printed three tables of constants widely used throughout the text.

w	water	I	inspired
g	gas	E	expired
b	blood	A	alveolar
a	arterial	L	pulmonary
v	venous	D	dead space
$\bar{v}$	mixed venous	T	tidal
c	capillary	br	branchial
t	tissular	B	barometric

Quantity	*Symbol*	*Example of unit*
Solubility coefficient, *e.g.*	α	$nmol \cdot cm^{-3} \cdot Torr^{-1}$
O_2 solubility in water	αw_{O_2}	$nmol \cdot cm^{-3} \cdot Torr^{-1}$
Capacitance (coefficient),	β	$nmol \cdot cm^{-3} \cdot Torr^{-1}$
e.g. capacitance of CO_2 in water	βw_{CO_2}	$nmol \cdot cm^{-3} \cdot Torr^{-1}$
e.g. capacitance of a gas species in a gas phase = $1 \cdot (RT)^{-1}$	βg	$nmol \cdot cm^{-3} \cdot Torr^{-1}$
Concentration, *e.g.*	C	$mmol \cdot L^{-1}$
concentration of CO_2 in mixed venous blood	$C\bar{v}_{CO_2}$	$mmol \cdot L^{-1}$
Difference of concentration, *e.g.*	ΔC	$mmol \cdot L^{-1}$
ΔC of O_2 between inspired and expired medium	$\Delta C_{I,E_{O_2}}$	$mmol \cdot L^{-1}$
Diffusion coefficient (diffusivity),	D	$cm^2 \cdot sec^{-1}$
e.g. D of O_2 in water	Dw_{O_2}	$cm^2 \cdot sec^{-1}$
Fraction of a gas species in a gas phase, *e.g.*	F	dimensionless
F of N_2 in inspired air	$F_{I_{N_2}}$	
Body mass	B	kg

Krogh's constant ($=D\cdot\beta$), *e.g.*	K	$nmol \cdot cm^{-1} \cdot sec^{-1} \cdot Torr^{-1}$
K of O_2 in water	Kw_{O_2}	
Extraction coefficient, *e.g.*	E	dimensionless
extraction coeff. of blood O_2	Eb_{O_2}	
Ventilatory frequency	fR	min^{-1}
Cardiac frequency	fH	min^{-1}
Partial pressures of O_2	P_{O_2}	Torr
Partial pressures of CO_2	P_{CO_2}	Torr
e.g. P_{O_2} in arterial blood	Pa_{O_2}	Torr
P_{O_2} in blood	Pb_{O_2}	Torr
Partial pressure of H_2O at temperature T′	$P^{T'}_{H_2O}$	Torr
Difference of partial pressure between two milieus, *e.g.*	ΔP	Torr
P_{O_2} between inspired air and alveolar gas	$\Delta P_{I,A_{O_2}}$	Torr
Rate of O_2 consumption	$\dot{M}_{O_2}$	$mmol \cdot min^{-1}$
Rate of CO_2 production	$\dot{M}_{CO_2}$	$mmol \cdot min^{-1}$
Electrical resistance (Ohm's law)	R	Ω
By analogy R may be a physiological resistance, as R between inspired and expired milieus	$R_{I,E_{O_2}}$	$Torr \cdot min \cdot mmol^{-1}$
G is a conductance $= R^{-1}$		$mmol \cdot min^{-1} \cdot Torr^{-1}$
Respiratory ratio	R	dimensionless
Without subscript R is the respiratory quotient in steady state and has a metabolic meaning, $\dot{M}_{CO_2} \cdot \dot{M}_{O_2}{}^{-1}$	R	dimensionless

but R between two milieus should be specified, *e.g.* R by comparing inspired and expired gas composition	$R_{I,E}$	dimensionless
Thermodynamic constant	R	$J \cdot mol^{-1} \cdot K^{-1}$
Ventilatory water flow rate	$\dot{V}_W$	$ml \cdot min^{-1}$
Ventilatory air flow rate	$\dot{V}g$ or $\dot{V}air$	$ml \cdot min^{-1}$
Inspiratory air flow rate	$\dot{V}_I$	$ml \cdot min^{-1}$
Expiratory air flow rate	$\dot{V}_E$	$ml \cdot min^{-1}$
Blood flow rate (cardiac output)	$\dot{V}b$ or $\dot{Q}$	$ml \cdot min^{-1}$
Systemic blood flow rate	$\dot{V}b_{sys}$	$ml \cdot min^{-1}$
Pulmonary blood flow rate	$\dot{V}b_L$	$ml \cdot min^{-1}$
Pulmonary resistance to blood flow	R_L	$Torr \cdot min \cdot L^{-1}$
Systemic resistance to blood flow	Rsys	$Torr \cdot min \cdot L^{-1}$
Tidal volume	V_T	L
Stroke volume	V_S	L
Blood pressure in pulmonary artery	P_{PA}	Torr
Blood pressure in aorta	P_{SA}	Torr
Thermodynamic temperature	T	K
Biological temperature	T′	°C

CHAPTER 1

The scope of respiration

Summary

Oxygen consumption and carbon dioxide production take place in the cell. Since O_2 comes from the surrounding medium, water or air, and CO_2 is rejected into it, the O_2 and CO_2 molecules have to be transferred through various barriers and compartments, either by diffusion over a short distance, or by convection, external breathing and circulation, over long distances. These are the fundamentals of the respiratory system. However, the field covered by the word respiration is not well delineated, and it should not be, because no apparatus, system or function of a whole, which is the organism, can be envisaged in isolation.

The physicochemical characteristics of the environments, mainly aquatic and aerial, are very numerous, variable and often hostile to life. Various strategies of survival are observed which can be classified into two categories: (1) the animals reduce their metabolism until propitious ambient conditions again prevail; (2) the animals develop ways of coping with new environmental situations and maintain their metabolism.

The respiratory system

Living organisms need energy. Most animals obtain energy from the catabolism of organic substrates which are derived from green plants using solar energy. Generally energy is produced from oxybiotic respiration, that is, for the animal physiologist, the use of oxygen and the production of carbon dioxide.

Cellular respiration requires that oxygen reach the vicinity of the cells at a partial pressure high enough to ensure its diffusion to the site of its utilization. Simultaneously, the cells must get rid of the carbon dioxide they produce, so that the CO_2 partial pressure and the acid–base balance (ABB) of the extracellular fluid are compatible with the life of the cells.

Only in protozoans and certain minute multicellular animals can respiration be sustained uniquely by a simple diffusive process. Organisms above a certain size have complex body plans composed of systems subserving particular functions. Oxygen must be transferred between the environment and the cells where respiration takes place through a series of structures: the breathing apparatus, the wall separating the ambient medium from the blood, the circulatory system, the wall between the blood and the fluid surrounding the cells, the cell membranes; carbon dioxide must make the reverse transfer. The transfer involves two convective systems, one for external breathing and one for blood transport, and two diffusion barriers, namely a water- or air-to-blood barrier, and a blood-to-cells barrier.

The disposition of the respiratory system varies greatly in detail. A special mention must be made of the insects, the most successful group of animals considering the overwhelming number of existing species. In insects, external respiration may be either only diffusive, particularly in some insects in which no ventilatory movements are observed, or by ventilatory convection of big tracheae and air sacs and by diffusion along a tracheolar system which subdivides itself progressively, the terminal tracheoles ending near the cells or even indenting them. The hemolymph which in insects usually does not contain any O_2-carrying pigment seems to play no important role in the transport of oxygen to the tissues, but could play a role in the transport of CO_2 whose capacitance in hemolymph, even without pigment, is not negligible.

Anyway, in all cases, the body is the seat of a net inwards flux of oxygen, the so-called O_2 uptake, and a net outwards flux of carbon dioxide, the CO_2 output. By an unhappy convention, these two vectorial quantities, the net influx of O_2 and the net efflux of CO_2, were given a positive sign, so that the respiratory quotient is normally positive. The intensity of these fluxes is quite variable. They may be increased in a given animal by one or two degrees of magnitude with various functions: nutrition and metabolism, temperature, exercise, sexual or reproductive activities. The double, oppositely directed transfers must be simultaneously satisfied through all these steps over a wide range of O_2 and CO_2 fluxes. There would be no point in having a working capacity of one step, diffusive or convective, of the respiratory system greater than the working capacities of the other various links of this system. It is sufficient that they are sized similarly to ensure the required O_2- and CO_2 fluxes. In this characteristic of the respiratory system, Weibel and Taylor (1981), Taylor *et al.* (1987) and Weibel *et al.* (1987) see an example of *symmorphosis*. The functions of the various structures or designs of the respiratory system must be regulated. If not, respiratory failure ensues, and the cells suffer asphyxia, the association of a lack of oxygen, hypoxia, and a superabundance of carbon dioxide, hypercapnia.

It took more than one century to arrive at the concept of 'the respiratory system' (see Keilin, 1966). Lavoisier (1743–1794) is usually credited with the demonstration that respiration is a chemical phenomenon in which oxygen is used and carbon dioxide is produced, a chemical process which he compared to the combustion of coal, animal heat being considered as analogous to the heat produced by burning fuel. Lavoisier had precursors: in the 17th century in Oxford, R. Boyle, R. Lower and J. Mayow came very close to discovering the true nature of respiration and the origin of animal heat. One will note here that the big question debated in the 18th century was the origin of animal heat, which is clearly observable only in birds and mammals. At a time when organic chemistry did not exist, the concepts of nutrition and metabolism were lacking, the concept of tissue and of the cellular constitution of organisms was unknown, it is obvious that Lavoisier and his contemporaries could not decide that respiration was basically a cellular process. However, had they turned their attention to lungless, bloodless animals, the site of the respiratory process would probably have been recognized sooner. Spallanzani (1729–1799), founder of the comparative physiology of respiration, observed that the preparations he examined – leeches, earthworms, insects, various fragments of the organisms studied *in vitro* – all use oxygen and produce carbon dioxide (1807). But the reports of these observations were overlooked for many years (see Keilin, 1966), and in the 19th century, four sites were successively proposed for respiration: (1) the lungs, (2) the circulating blood, (3) the capillary blood, and (4) the cells and tissues. It is Pflüger who is credited with the statement that the elementary process of respiration takes place in the cell (1872). Note that between the end of the 18th century and Pflüger, two fundamental fields of modern sciences were established, namely organic chemistry and the cellular theory.

Since the oxygen is used and the carbon dioxide is produced in the cells, it is obvious that the organism is the seat of a net inwards oxygen flux and of a net outwards carbon dioxide flux which imply various steps in the transfer of O_2 and CO_2. This is stated clearly enough by Paul Bert (1870) although the true nature of cellular respiration had to await the discovery of the respiratory pigments, the histohematins of MacMunn (1886), renamed cytochromes by Keilin (1923), and the modern development of mitochondrial chemistry. Chapters 4 and 6 will illustrate the concept of respiratory system.

Anaerobic respiration

If an increase of the net influx of oxygen is insufficient to meet the needs of energy expenditure, the animal can still produce some energy by using the reserves represented by its stores of phosphagens and adenosine triphosphate, or by turning for short periods to fermentation, which is much less efficient than oxybiotic respiration, or by both. But these processes of energy production cannot continue for long, except in some parasites, or in more complex animals under conditions of low energy requirement, such as various forms of wintering over or estivation. Sooner or later, the animal must return to oxybiotic energy production, and, in order to reconstitute its initial body composition, it must pay an oxygen debt. Thus respiration includes anoxybiotic fermentative energy production as well as oxybiotic, these two processes being to a certain extent complementary.

Whether the energy production is oxybiotic or not, substrate catabolism produces acids, mainly carbon dioxide and carbonic acid in respiration, various organic acids in fermentations. Respiratory physiology must take into account all the situations of acid–base balance (ABB) in the various body fluids, and consequently those related to the cutaneous, branchial and renal ionic exchanges and their regulation, to the extent that these processes are connected with the respiratory or fermentative energy production of the animal. Dealing with the regulation of ABB as a part of respiratory regulations is necessary because acid–base imbalance may in turn disturb the respiratory processes, and even contribute to regulating them.

The field covered by the word 'respiration' is by no means clearly circumscribed, nor should it be. This is true of any field of physiology. Rigid boundaries mean the setting of artificial limits, isolating certain functions out of context in the complexity of the living being, all of whose functions and regulations operate and interact simultaneously. Since, however, it is impossible to grasp immediately the whole of life, one is obliged to concentrate on one or a few biological functions at a time. Respiration as here delimited indeed goes beyond what is commonly understood by the word, yet it is still an arbitrary and restrictive approach, which, for the time being, cannot be avoided.

Examples of interference between respiration and other functions will be examined. From these, it will appear that the preeminent property of respiration is the delivery of oxygen to the cells. Where there is conflict in the simultaneous satisfaction of various needs, the oxygenation of the body is always preserved at the expense of the other functions, which are subordinate: thus there is a hierarchy of functions. But the hierarchy may differ for different members of the animal kingdom; what may be true for aquatic animals may not be true for terrestrial animals, above all the homeotherms.

The environment, stress, survival

An animal's respiration is regulated not only in function of what is happening inside its own body, but, obviously, of what medium, water or air, it is living in, and of the stresses that the environment may impose (Bligh *et al.*, 1976). Necessarily the organisms are adapted to their environment, or they would not exist. Chapter 3 will take up the so-called respiratory properties of the medium breathed, that is, those concerning O_2 and CO_2. In Chapter 9, most of the physicochemical properties of water and air, all of which have a direct or indirect bearing on respiratory functions will be contrasted and will show that, in a general way, animals living in air have anatomical and physiological features differentiating them from aquatic animals.

The physicochemical properties of the environment vary in space and time. The chronological variations follow various cycles: tidal, circadian, lunar, seasonal. The environmental factors, modified by topographical and chronological traits, define a livable environment, a life-supporting space (fig. 1.1). They constitute a complex geometrical space in n-dimensions in which life evolved throughout millions of years. The boundaries of this space are by no means well defined, because the envelope of this environment is species-specific. Within the n-dimensional livable space,

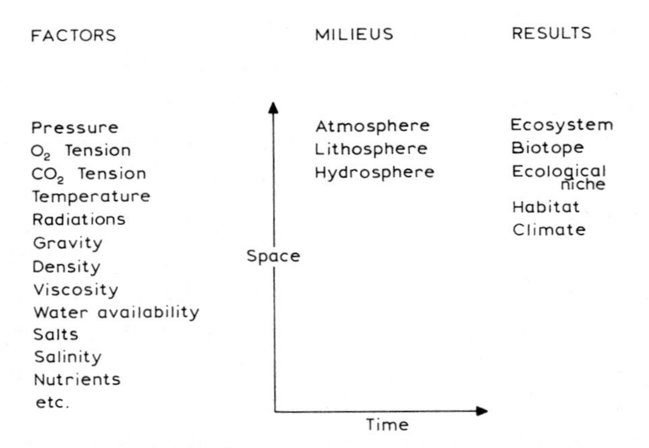

Fig. 1.1. Ambient factors which change in time and space in the various milieus, water, soil and air, to result in a given environment.

the physicochemical conditions are tolerable, and there are zones where living beings can prosper and develop relatively easily. Beyond the boundaries, life is not possible. Just inside, the conditions are tolerable for some time and the organisms are strained. Various strategies of survival are then observed which may be classified into two categories: (1) the animals possess means of coping with new situations and maintain their metabolism; (2) the animals reduce their metabolism until propitious ambient conditions again prevail. In all cases, the organism facing hostile conditions can survive thanks to some adaptations which can be defined as changes minimizing the physiological strain which results from a stressful environment (Farhi, 1987). Some examples will illustrate this defense against unpropitious environmental factors.

(1) Against cold, some animals try to increase their energy production through shivering or hormonally controlled non-shivering thermogenesis, or both, or try to increase their thermal insulation by vasoconstriction and eventually by increasing the thickness of the integument, a delayed process of adaptation. Against heat, animals use various tricks to increase their heat loss. Other animals may regulate behaviorally against cold or heat. They may search for a better place in the world, either locally by finding a more favorable microenvironment (galleries, caves), or by migrating to more hospitable climates.

(2) If the ambient conditions are not favorable (heat, cold, water scarcity, lack of food) then the animals can react *in loco* and find a place to

estivate or to winter over, *i.e.* to enter into a state of dormancy requiring a decreased energy expenditure (Hochachka and Guppy, 1987). This phenomenon is observed in many animal groups including the hibernating mammals. The blood values of P_{CO_2} and of the acid–base balance during dormancy as compared to active state are markedly altered. For example in the land snail *Otala lactea* at 25 °C, pH and P_{CO_2} values are respectively 7.86 and 11 Torr on the active animal and 7.21 and 63 Torr in the dormant animal in a dry atmosphere (table 4.2, p. 38). Analogous observations have been made in the active and cocooned African bullfrog at 25 °C (Loveridge and Withers, 1981). Also the pH and P_{CO_2} values of the arterial blood of the European hamster *Cricetus cricetus* are respectively 7.40 and 45.3 Torr in the active hamster with a body temperature of 37 °C, and 7.57 and 36.1 Torr when the animal is hibernating at 9 °C (table 4.2, p. 38). These two last values during hibernation indicate actually a very marked hypercapnic acidosis, since in the blood sampled at 37 °C and cooled down in a syringe to 9 °C, the pH and P_{CO_2} values would be respectively 7.84 and 10.1 Torr (table 4.2). For Malan (1973, 1986) and for Barnhart and McMahon (1987) the respiratory acidosis observed in dormancy may be a factor responsible for the enormous fall of the metabolic rate observed in this circumstance.

(3) At the extreme, when the ambient conditions are not favorable for survival, some animals can enter into a state of latent life, or cryptobiosis, in which the energy expenditure falls practically or really to zero, as is observed in some relatively simple invertebrates: rotifers, nematodes, tardigrades (Keilin, 1966; Womersley, 1981) and in various cells or tissues (Mazur, 1984).

Obviously all species we observe today have triumphed over all the dangers to their survival which may have occurred in the past. No doubt, some species were at a time *almost* extinct; but it is sufficient that some individuals persisted in a remote niche to ensure the perpetuation of the species and its eventual flourishing development in favorable environments. It may happen that some adaptive mechanisms are never solicited during a life time because the ambient conditions never reach a stressful threshold. These mechanisms are nonetheless very important because they may come into play in conditions which threaten life and save some individuals and the species. A pertinent example is to be found in mammalian physiology. It is the resistance of neonates to asphyxia at birth. During labor in many mammals there occurs a physiological asphyxia which the neonate can withstand. Their resistance to asphyxia, which in particular spares the brain (p. 112), may play a role once in a life time; anyway it is fundamental since without it most births would be stillbirths.

That animals are adapted to their environment may seem to be an obvious, indisputable kind of statement. However, the mutual harmony of life and environment seems so wonderful that L. Henderson (1913) was able to write: "Darwinian fitness is compounded of a mutual relationship between the organism and the environment. Of this, fitness of environment is quite as essential a component as the fitness which arises in the process of organic evolution; and in fundamental characteristics the actual environment is the fittest possible abode of life." We meet here an exemplary intellectual phenomenon in the mind of a biologist. A biologist is fascinated by the phenomena of life, by the behavioral and physiological mechanisms of living beings by which they perpetuate themselves in their environment, by the necessary and amazing adaptations of animals and plants to their surroundings. However, living beings presumably really adapt to their environments, and not the other way around. Certainly there may be in the universe some environmental conditions which are different from those of our planet, where life as we know it evolved. But it remains difficult to reason that an environment has evolved in order to be fitted to the living. I feel that it is the other way around: living beings are adapted to an environment or they do not exist.

CHAPTER 2

Phenomena, mechanisms, regulations

Summary

To say that living beings are adapted to the environment is a truism. How could it be otherwise? Cells, tissues, organs, apparatuses and systems form organisms which exhibit some living phenomena. Mechanistic physiology deals with the qualitative, semi-quantitative and quantitative relations between the phenomena. Their regulations are such that the relations complement each other and environmental constraints are overcome.

For the respiratory system a few examples will be given. Some regulations take precedence for the animal's survival over others; certainly the correct steady delivery of oxygen to the cells has priority over many other regulations, including the regulation of the CO_2 clearance and of the acid–base balance. Thus there exists some hierarchy among regulations, because only rarely can all functions be correctly regulated at the same time.

Whatever the unit of life under study, a cell, a tissue, an organ, an apparatus, a system, an organism, a colony, a biosystem, there are four interwoven matters to examine: (1) the structures, molecular, subcellular, cellular, supracellular: chemistry, morphology and anatomy; (2) the activities exhibited, the manifestations of life: phenomena; (3) the explanatory mechanisms, the qualitative, semiquantitative or quantitative relations between the phenomena: mechanistic physiology; (4) the regulations of the mechanisms such that they complement each other and provide for adjustment to environmental constraints: regulatory physiology.

Phenomena and mechanisms

The differences between phenomenological, mechanistic and regulatory physiologies are rarely clear-cut, and may be confused with each other. An illustration can be taken from respiratory physiology. If respiration is to be viewed as a system for O_2 and CO_2 transfer involving various steps of diffusion and convection, the transfer through all these steps is governed by some quantitative relations. In steady state, all these relations, expressed as equations, must be simultaneously satisfied. There is no exception to these equations as long as models to which the equations pertain coincide with the phenomena exhibited by real animals. To explain the O_2 transfer from air to the tissues, one can use these equations which relate the net O_2 flux, $\dot{M}_{O_2}$, the difference of partial pressure of oxygen between two media, ΔP_{O_2}, and a factor of proportionality, the resistance R (Fenn, 1964).

Between air and alveolar gas, we have the equation

$$\Delta P\text{I, A}_{O_2} = R\text{I, A}_{O_2} \cdot \dot{M}_{O_2} \tag{2.1}$$

between alveolar gas and arterial blood

$$\Delta P\text{A, a}_{O_2} = R\text{A, a}_{O_2} \cdot \dot{M}_{O_2} \tag{2.2}$$

between arterial and venous blood

$$\Delta P\text{a, v}_{O_2} = R\text{a, v}_{O_2} \cdot \dot{M}_{O_2} \tag{2.3}$$

These equations have the form of Ohm's law

$$V = R \cdot I \tag{2.4}$$

where V, R and I mean respectively the electrical potential, the electrical resistance and the current intensity.

Since air, alveolar gas, arterial blood, and venous blood are in series, eqs. (2.1), (2.2) and (2.3) in steady state may be combined:

$$\Delta P\text{I, v}_{O_2} = R\text{I, v}_{O_2} \cdot \dot{M}_{O_2} \tag{2.5}$$

in which

$$\Delta P_I, v_{O_2} = \Delta P_I, A_{O_2} + \Delta P_A, a_{O_2} + \Delta Pa, v_{O_2} \quad (2.6)$$

$$R_I, v_{O_2} = R_I, A_{O_2} + R_A, a_{O_2} + Ra, v_{O_2} \quad (2.7)$$

At sea level, the pressure of oxygen, P_{O_2}, is $\simeq 150$ Torr; at the top of Mount Everest it is $\simeq 42$ Torr. Thus, if O_2 consumption, $\dot{M}_{O_2}$, is to be the same at the two elevations, eq. (2.5) shows that R_I, v_{O_2} must decrease by the same proportion as the decrease of P_I, v_{O_2}. One knows the components of the resistance of eqs. (2.1) and (2.3), that is the reciprocal of the product of the flow rate of gas (2.1) or of blood (2.3) times the O_2 capacitance in gas (2.1) or in blood (2.3). The intermediate components of eq. (2.2) are more complex, but are understood semiquantitatively (Dejours, 1981).

Thus it is possible to figure out how oxygen can be transferred from inspired air to venous blood at sea level and at very high altitude (Dejours, 1982). For example, alveolar ventilation must increase at high altitude; there is also an automatic increase of the blood oxygen capacitance because of the sigmoid form of the blood's O_2 absorption curve; and in spite of the difficulty of transferring O_2 by diffusion with a low O_2 pressure head, the resistance R_A, a_{O_2} must remain low. None of these conditions can be avoided; if they had not been fulfilled, it would not have been possible for humans to climb Mt. Everest breathing air. They are necessary and we must be able to explain how an equation can be satisfied.

This to a certain point is very gratifying; however, nothing is said about the regulatory mechanisms. Some are known, some are not. There is no question that the rich meaning of an equation, the beauty of some necessary relations, the fact that one can predict how a factor varies as a function of another, are quite satisfying; but it is disappointing if little is known about the regulations underlying these relationships.

There is thus a permanent confusion between mechanistic physiology and regulatory physiology. It is difficult sometimes to distinguish between the two, as it may be hard to differentiate mechanistic from phenomenological physiology. At many places, this book will not go farther than describing phenomena, and problems of regulation will not really be dealt with either because they are completely unknown, or because the answers offered are too speculative. Anyway, even if with regard to some functions one may speak of the nature of the regulation, one can never pretend to know the whole story. As soon as some regulations are understood, then other questions about the intimate mechanisms of these regulations are raised.

A way of getting around this is to introduce a black box (and we introduce more than we think); it contains an unknown world. Some thirty years ago, I extensively and purposely represented the arterial chemoreceptor bodies as black boxes; at that time, very few physiologists were concerned with the intimate mechanisms occurring inside. Today several tens of teams are trying to decipher them. The black box has one advantage: it permits one to go on, to look beyond, and to solve some further problems. But by the same token, it presents a danger: the success of the black box hypothesis is sometimes so outstanding that one just forgets in the meantime that many unknowns are contained in a very black box; in so doing, it is no longer a black box, but a magic box.

What is a regulation?

The concept of regulation of animal functions is certainly very old, although its systematic study was not fully developed until Claude Bernard. Claude Bernard himself rarely used the word 'regulation', but more often the terms 'compensation' and 'balance'. Since then, the concept of regulation has been considerably developed by biologists, physicists, engineers, economists and sociologists. It is nowadays an individualized part of science, with its own vocabulary, its specific graphical symbolism and mathematical formalism.

Simple examples of regulation are found in human civilization. It is easy to understand an all-or-nothing regulatory system, such as a refrigerator, a regulated domestic heating system, or a thermostatted bath. To define some fundamental terms, I will take the example of an oxystat, an apparatus for regulating the O_2 tension in a given medium.

Let us suppose we have a tank of sterile water thermostatted at 20 °C in *equilibrium* with air bubbling through it and at a barometric pressure of 760 Torr; then the O_2 tension in the air and the water is 156 Torr. Animals put in this water will use dissolved oxygen and the value of P_{O_2} will fall. As soon as P_{O_2} starts falling, there will be a net O_2 flux from air to water. Eventually, depending on the arrangement of air bubbling and the O_2 consumption of the animals, a new steady state may be reached at an oxygen pressure of *e.g.* 50 Torr, in which case the O_2 flux from air to water will be equivalent to the O_2 consumption of the animals, $\dot{M}_{O_2}$.

But we may want to avoid this hypoxic condition and to keep water P_{O_2} at, let us say, 130 Torr. We can increase the bubbling manually by trial and error and arrive at a P_{O_2} of about 130 Torr, but it would require dexterity and exhausting day and night monitoring. Instead we may use an automatic oxystat (fig. 2.1). In the water, the controlled system (1 of fig. 2.1), we want to keep P_{O_2}, the controlled signal (2), at 130 Torr. The actual O_2 tension in the bath is sensed by an O_2 electrode connected to an O_2 meter (3). The meter output (4) feeds back to an error detector (5) set at the desired value of P_{O_2} (130 Torr), the set point (6). If, for instance, the water P_{O_2} goes higher than the desired P_{O_2}, the error detector transmits an all-or-nothing signal (7) to a solenoid valve (8) which closes the line coming from an O_2 tank or an air pump. Consequently since the animals use dissolved oxygen, P_{O_2} decreases in the water bath, the controlled system (1). As soon as the water P_{O_2} falls to the set point (130 Torr) of the error detector, the solenoid valve opens again (or the air pump is switched on) and P_{O_2} will soon rise. The oxygen thus admitted, $\dot{M}_{O_2}$, is the controlling signal (9). The terms to retain are: controlled system (1); controlled

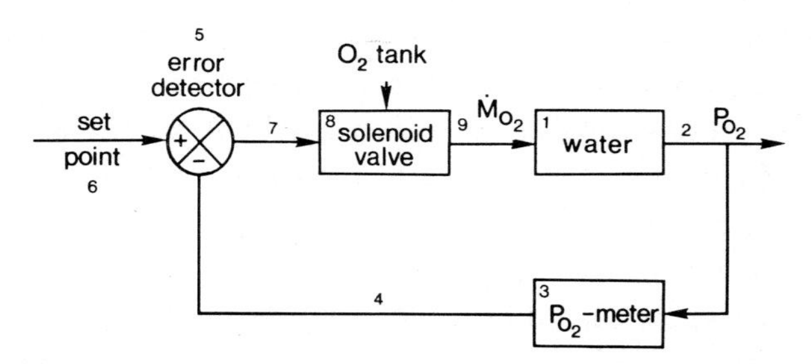

Fig. 2.1. Block diagram of a P_{O_2}-stat. The controlled system is a volume of water (1) at constant temperature, *e.g.* an aquarium with some fish, through which a gas mixture (*e.g.* air) is to be bubbled or not. The regulated signal is the water P_{O_2} (2) monitored by an immersed P_{O_2} electrode and P_{O_2} meter (3). The meter output (4) feeds back to an error detector (5) set at the desired P_{O_2} value (6), the set point, in this case 130 Torr. If the water P_{O_2} value is lower than the desired value, 130 Torr, the error detector transmits an all-or-nothing signal (7) to a solenoid valve (8) which opens to let in pure O_2 or an O_2-containing mixture with $P_{O_2} > 130$ Torr (as air at sea level). Thus, some O_2, $\dot{M}_{O_2}$, flows into the gas bubbling through the water, where it is mixed. $\dot{M}_{O_2}$ (9) is the controlling signal. The O_2 dissolves in the water and increases P_{O_2}. As soon as the water P_{O_2} value reaches the set point of the error detector, the solenoid valve closes. Then, as O_2 is used by the fish, P_{O_2} falls in the water and the process of O_2 bubbling will again be started up.

signal (2); sensor (3); error detector (5); set point (6); controlling signal, $\dot{M}_{O_2}$ (9); they are clearly defined.

In this system, the water (1) is subject to a double regulation, of temperature and O_2 tension. A pH-P_{CO_2} regulation may be added. To have a regulation a regulatory loop is needed. A system without feedback is not looped and is not regulated. Any disturbance will change the regulated signal. Without a regulatory arrangement, the water P_{O_2} of our example would have fallen to 50 Torr; furthermore this value would not have been steady since it would have changed as a function of the animals' activity (exercise, circadian rhythms).

A water system in which P_{O_2}, pH-P_{CO_2} and temperature must be regulated is already complex; for example all regulations must be recalibrated if one wishes a bath thermostatted at a different temperature, or if the water composition changes, because these changes imply some concomitant changes of the O_2 solubility and the P_{CO_2}-pH status in the water. However, a water bath with these three regulated variables: temperature, P_{O_2} and P_{CO_2}-pH, works very well.

The foregoing example illustrates the elements of an all-or-nothing regulatory system. However, many man-made regulations and most physiological regulations are more complex. In particular our example makes use of an all-or-nothing controlling signal, $\dot{M}_{O_2}$. In reality many regulations use a progressively increasing or decreasing controlling signal which may be proportional to the difference between the actual value of the controlled signal and the value of the set point. These *proportional regulations* are often encountered in the living world. A proportional regulation is better for avoiding 'hunting', which consists of oscillations of the regulated signal around the set point. In the foregoing example, if the controlling signal, $\dot{M}_{O_2}$, is too high, the water bath will be hyperventilated and the O_2 pressure may swing up after the air entry is closed off. On the other hand, if $\dot{M}_{O_2}$ is too low, it may happen that the O_2 pressure falls very low after the air entry is switched off and that it takes a long time to return to the set point. Of course, we can fix the magnitude of the controlling signal to avoid broad hunting; by doing so, we use another regulating device, namely our own mind. This is true for all man-made regulations, even the proportional ones. We do this by experience, by trial and error.

The complexity of respiratory regulations

In living organisms, the regulations are complex. They are very numerous; they are intricate, and most of the time they interact. It is not very easy to say what are the controlled systems, what are the regulated signals, where are the error detectors, what are the set points and the controlling signals. In other words: What is regulated? How is it regulated?

Respiratory regulations require that O_2 tension be high enough in the vicinity of the cells to ensure its diffusion to the site of its utilization in the cells, that the CO_2 tension not be too high, and that the acid–base balance of the body fluids be proper. In complex organisms, as noted, many steps are disposed serially, ensuring the adequate fluxes and pressures of O_2 and CO_2. Each of these steps must be simultaneously regulated. If one fails, the O_2 delivery to the cells is impaired.

Let us consider first the problem of delivering an adequate amount of oxygen to the cells in a complex organism. The following regulations are required:

(1) Regulation of external respiration, including the structures ensuring breathing and the transfer of oxygen from the external medium to the blood; here the controlled system is the arterial blood and the controlled signal is the arterial P_{O_2}. We will see later (p. 65) the complete regulatory loop involved in the regulation of Pa_{O_2} and the eventual role of carbon dioxide.

(2) Regulation of the O_2-carrying properties of the blood. Here the controlled system is the blood, the regulated signal is the concentration and O_2 affinity of hemoglobin. The kidneys, mainly the juxtaglomerular cells, sense their own local oxygenation, and apparently play two other roles: detection of the error of their own oxygenation and delivery of proper amount of erythropoietin which acts on the controlling system, the bone marrow cells, which in turn increase or decrease their activity, that is their rate of erythrocyte and hemoglobin formation; the controlling signal is the rate of erythrocyte number delivered by the bone marrow to the blood.

(3) Regulation of the blood circulation. This is particularly complex since we have two subsystems: the ‘high pressure’ system (namely the left ventricle, the systemic arteries, the arterioles and their sphincters) and the ‘low pressure’ system (the systemic veins, the right heart, the pulmonary vessels, the left auricle).

(a) Regulation of the 'high pressure' system. The controlled system is the blood, the controlled signal is the arterial blood pressure; the sensors are the baroreceptors; the error detector is located in the brain stem; the controlling signal is the activity of the neurovegetative centers of the brain stem which modulate the activity in the parasympathetic and sympathetic efferents to the heart and vessels.

(b) The 'low pressure' system, in particular the venous return and the pulmonary circulation, are also regulated. The regulation of the blood circulation is more complex since other controls, such as the control of blood volume by hormones, may intervene.

(4) Regulation in organs and tissues, which are interposed between the 'high pressure' and 'low pressure' systems and which oppose resistances to the flow of blood. The controlled system is represented by the cells, but since organs and tissues are made of various types of cells, it is topically poorly defined. Most probably the main controlled signal is the local oxygenation, or any factor closely related to the state of the local oxygenation. The cellular error detector and the controller should be looped together. Some chemical factors seem to act directly by diffusion on the smooth muscles controlling the local blood flow and the patency of capillaries. Such mechanisms decrease the resistance to blood which is under high pressure in the arteries. This scheme seems to be valid at least for the hyperemia (increase of patent capillaries, increased blood flow), observed in exercising muscles.

The speed with which regulations come into play is variable. The ventilatory flow and the blood flow, the vascular resistances may change very quickly, within a few seconds, whereas it may take days or weeks to observe changes in the structures of the breathing apparatus, the size of the heart, or the erythrocyte and hemoglobin concentrations, all of which require biochemical syntheses.

Hierarchy of regulations

Generally a regulation which favors O_2 delivery to the cells favors the CO_2 clearance from the cells and aims at balancing the acid–base status. For example in exercise, the organism's O_2 uptake from the ambient medium may be increased 10–20 times or more, as well as the CO_2 clearance. The above-mentioned regulations favor O_2 *and* CO_2 transport. However, at high altitude, for a given O_2 consumption, a regulation which permits the O_2 delivery to the tissues unbalances the CO_2 clearance and the acid–base balance. Also, the regulation of body temperature can affect the CO_2 clearance and the ABB (p. 102). Other examples, as we will see (Chapter 7) can be given. This illustrates the fact that the regulatory mechanisms often interact in directions which are not favorable for all the controlled variables (the regulated signals in terms of control theory). In the case of conflict of the regulatory mechanisms, it is generally possible to recognize that one controlled variable is preferentially regulated. There exists thus some *hierarchy* of the regulations, because only rarely can they all be optimally regulated at one time. One of the aims of this book is to show that the regulation of oxygenation in the body, namely of the tissues, or organs, or of certain organs as the brain (p. 109), is often preeminent, while some other regulations, *e.g.* regulations of CO_2 clearance and of acid–base balance, are subordinate.

Biological regulations are not, however, fixed once and forever. During development they are progressively installed; they reach their optimal activity at different ages. Then they lose their efficiency with age. Here we meet the problem of ontogeny and aging of regulations. On the other hand, regulations can be modified by differing demands on the organism. If the environment changes (high altitude, ambient temperature), if the animals are sedentary or active, the regulations are modified. Regulatory mechanisms may be 'trained' and may adapt so that they intervene at a faster rate and to a broader extent, or at a different point allowing better performances. Animals are endowed genetically with a minimum of regulations without which they could not exist; but regulations may be reinforced epigenetically by new demands, by training or conditioning. An obvious example is muscular exercise in which the quality of the performance depends on long-term training, starting in childhood. This example is here of a particular interest since some exercise performances depend, among other things, on the quality of the respiratory regulations.

CHAPTER 3

Environments

Summary

Concerning O_2 and CO_2, water and air have completely different properties. Waters differ extremely and the physicochemical properties of a given water must be known in detail in order to calculate how O_2 and CO_2 partial pressures and concentrations will change when O_2 and CO_2 are taken from or added to the water. When one speaks of CO_2, the acid–base balance of the water is automatically involved.

Atmosphere is a relatively simple milieu, *i.e.* little information (temperature, barometric pressure) is necessary to predict the change of tension and of concentration which takes place when O_2 is consumed and CO_2 produced.

The soil is a very special environment which may be treated as an aquasphere when it is completely flooded, or as a confined atmosphere when it is dry and without free channels to open air. Partially wet soil must be considered as a porosphere whose exact properties depend on the relative quantity of the soaking water and its chemical nature.

When all factors of the environmental media are taken into account, composition, temperature, humidity, pressure, time variations (tidal, circadian, lunar, circannual), one is led to the concept of climates, or microclimates, of ecological niches which are considered in detail in books of ecology (Odum, 1971). In these vast fields of climates, three groups may be singled out: the aquatic environment, the atmospheric environment and the soil environment.

Before studying those properties of the milieus which have a direct relation with oxygen and carbon dioxide, some general principles of respiratory physiology must be recalled. In particular the concept of capacitance, which has the dimension of a solubility, but which applies to any kind of milieu: water, gas, blood, *etc.* will be reviewed in detail. Indeed it is because the O_2 capacitance is much higher in air than in water, whereas the CO_2 capacitances are similar in both milieus, that the respiration of air breathers is so different from that of water breathers and that full homeothermy could originate in air breathers.

Capacitance coefficient

The solubility of a gas in a liquid is simple to visualize; the dimension of a solubility coefficient is (amount of substance)·(volume)$^{-1}$·(pressure)$^{-1}$, for example the unit can be mmol·L^{-1}·Torr^{-1}. The concept of capacitance is an extension of that of solubility; it applies to the capacity of gas and liquids, such as various waters and body fluids, to load or unload a given quantity of O_2, CO_2 or any other gas for a given change of partial pressure (Piiper *et al.*, 1971). Dimensions and units are the same for the solubility and capacitance coefficients.

The coefficient of capacitance of a gas x (O_2, N_2, etc.), β_x, in any medium is defined as follows:

$$\beta_x = \frac{\Delta C_x}{\Delta P_x} \tag{3.1}$$

ΔC_x being the change of the concentration of x corresponding to the change of its partial pressure, ΔP_x. If the relation between C and P is not linear, the equation must be written:

$$\beta_x = \frac{dC_x}{dP_x} \tag{3.2}$$

Waters

Any water whose temperature and salinity are known has a given solubility coefficient for O_2. There is also a solubility coefficient for CO_2 (fig. 3.1, left), but in most waters the relation between the total concentration of CO_2 and its partial pressure is more complex. If the water is distilled, or does not contain any buffer system, all the CO_2 is dissolved, whatever its partial pressure. But if the water contains some buffer (carbonates in many fresh waters; carbonates and, to a lesser extent, borates in sea water), its capacity to fix CO_2 is variable, because to the quantity of CO_2 which goes into solution may be added some quantity which is chemically bound. For example at high pH, a carbonated water contains some carbonate and some bicarbonate, the proportion between the two depending on the pH and the nature of the water. Carbonate can fix a proton according to the reaction:

$$CO_3^{2-} + H^+ \rightleftharpoons HCO_3^- \qquad (3.3)$$

Then, when a given quantity of CO_2 is dissolved in the water, some is hydrated as H_2CO_3, which immediately dissociates into a bicarbonate ion and a proton. The proton is picked up by CO_3^{2-} to give more HCO_3^-, and more carbonic acid can be formed and dissociated. These reactions will go on as long as some CO_3^{2-} is available. When all CO_3^{2-} has been transformed into HCO_3^-, then the forward reaction (3.3) stops. If then more CO_2 is added to the solution it will be dissolved and the slope of the relation C_{CO_2} *vs* P_{CO_2} will be the same as that of unbuffered water (fig. 3.2). The most important point here is that in normal, well-aerated buffered water, and that means most waters, the pH value is high and some buffer systems are present, so that the water breathers breathe generally in a water with a capacitance coefficient of CO_2 much higher than its solubility coefficient.

If the animals breathe water permanently, the duration of the water's contact with the respiratory surfaces is finite. By implication, the kinetics of the reactions concerning CO_2 loading should be high. Wright *et al.* (1986) measured the pH of the interlamellar water and reported that as water flowed over the gills, its pH significantly decreased. This implies that the formation of H_2CO_3 is rapid, presumably because of the presence of carbonic anhydrase on the water surface of the gills. One must also consider that the branchial filaments are lined by a layer of still water, the unstirred layer (USL). This USL water is permanently in contact with the mucus which covers the gills and presumably is the seat of the CO_2 loading of respired water. In the crabs which have been studied, carbonic anhydrase is in its highest concentrations in the gill tissues, a situation which presumably facilitates the CO_2 excretion (Burnett and McMahon, 1985; Burnett *et al.*, 1985).

Gas phase

The capacitance of a gas in a gas phase may be derived from the gas law

$$PV = MRT \qquad (3.4)$$

in which P stands for pressure, V for volume, M for quantity of substance, *R* for the thermodynamic constant and T for absolute temperature.

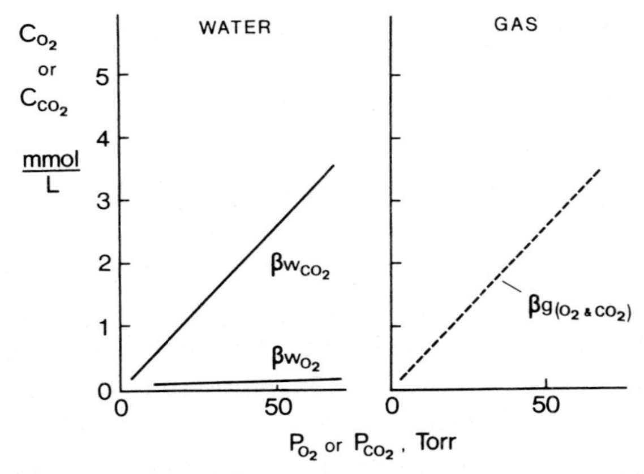

Fig. 3.1. Concentrations of O_2 and CO_2 in distilled water and in air at 15 °C plotted against P_{O_2} and P_{CO_2}. The slopes in water and in air correspond to the O_2 and CO_2 capacitance coefficients of these media. For distilled water and carbonate-free waters, capacitance coefficients are synonymous with solubility coefficients. Note that O_2 and CO_2 capacitance coefficients of a gas phase are identical (from Dejours, 1981).

In a gas phase, as long as the gases are nearly perfect and the pressure not very high, the case in the normal conditions of life, each gas x may be considered individually, and one may write

$$P_x V = M_x RT \tag{3.5}$$

Rearranging the equation

$$\frac{M_x}{V} = C_x = \frac{1}{RT} P_x \tag{3.6}$$

in which C_x is a concentration (amount of substance) $(\text{volume})^{-1}$. Figure 3.1 (right) shows the relationship between C and P at 15 °C. From eqs. (3.1) and (3.6)

$$\frac{\Delta C_x}{\Delta P_x} = \beta_x = \frac{1}{RT} \tag{3.7}$$

β_x, the coefficient of capacitance of the substance x, is valid for any gas and does not depend on altitude; it depends only, and to a small extent, on the absolute temperature (fig. 3.3). This figure shows also that the solubility coefficients of O_2 and CO_2 in an aqueous medium are much more dependent on temperature than the capacitance of a gas in a gas phase. Since β_x in gas phase is independent of the gas species, it is unnecessary to have a specific subscript; the symbol β_g may be used for all gases.

Other media

The concept of capacitance is valid for any medium (see p. 20). But in blood the relations between the concentrations of O_2 and CO_2 and their partial pressures are not linear. When arterial blood has to be compared to venous blood one has to use, *e.g.* for oxygen, the ratio

$$\beta a,v_{O_2} = \frac{\Delta Ca,v_{O_2}}{\Delta Pa,v_{O_2}} \tag{3.8}$$

which is the slope of the line joining points a and v. This can be called the effective arterio-venous capacitance coefficient of oxygen (Dejours, 1981). If arterial or venous blood or both change, this coefficient will also change.

The term 'capacitance' is used since something other than the solubility of physics is meant. It can be viewed as an 'effective solubility', as different from 'physical solubility'. In waters and in body fluids the 'effective solubility' is the sum of two quantities: (1) the quantity which is dissolved and can be calculated using the physical solubility coefficient; (2) a quantity which is chemically bound. In a gas phase, the quantity is directly proportional to *R*T; it is really nothing else than a dissolved amount of substance, but this word is not commonly used for gas in gas phase. For these various reasons, the word 'capacitance' is used to express in all media the ratio dC_x/dP_x.

Waters

The main fact regarding oxygen and carbon dioxide in water is that the solubility coefficient for O_2 is much smaller than the solubility, or capacitance coefficient for CO_2. As we will see, the consequence is that in water breathers the CO_2 tension is a fraction of a torr, or at most a few torrs above the ambient P_{CO_2}, whereas the O_2 tension is many tens of torrs below its value in the ambient water. Quite often carbon dioxide in water

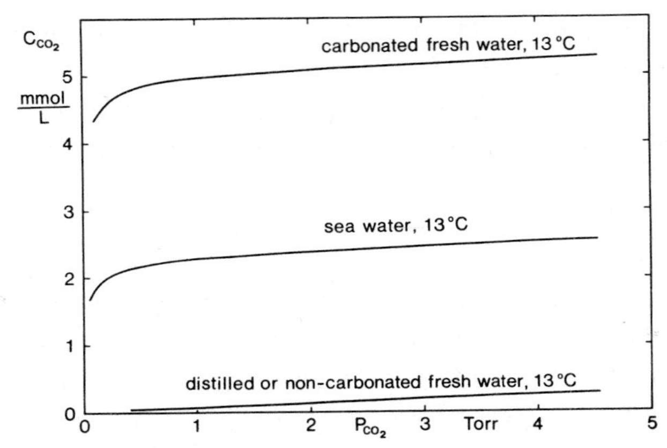

Fig. 3.2. Carbon dioxide absorption curves of different types of water (from Dejours *et al.*, 1968).

is buffered so that its capacitance coefficient is much higher than its physical solubility coefficient. Figure 3.2 shows CO_2 absorption curves of different waters: unbuffered water, sea water, and one example of carbonated water. Some waters are much more carbonated (see Livingstone, 1963), so that if the P_{CO_2} and the proton activity are not very high, their capacitance coefficients may be several hundred times higher than the physical solubility coefficient.

Another important property of CO_2 and O_2 in water is that their capacitance coefficients decrease markedly with the increase of temperature (fig. 3.3), whereas the energy metabolism and the respiratory requirement of the water-breathing animal increase with a temperature rise.

All kinds of brackish waters exist: estuarine water, and inland seas such as the Baltic sea (Harvey, 1957). To each salinity (and temperature) correspond known capacitance coefficients, but it is the change of salinity by its action upon ionoregulation which may be the most important factor for the physiology of animals.

Briny waters are generally very highly buffered. They are found in tropical and equatorial lakes and swamps (Livingstone, 1963).

The physicochemical properties of a body of water are usually not uniform. Near the surface O_2 and CO_2 may be equilibrated with air, but in deep water this is not generally the case. In ponds and swamps the bottom water is often very asphyxic. For example, in vegetation-covered lakes of the southeastern United States where two salamander species live, Siren

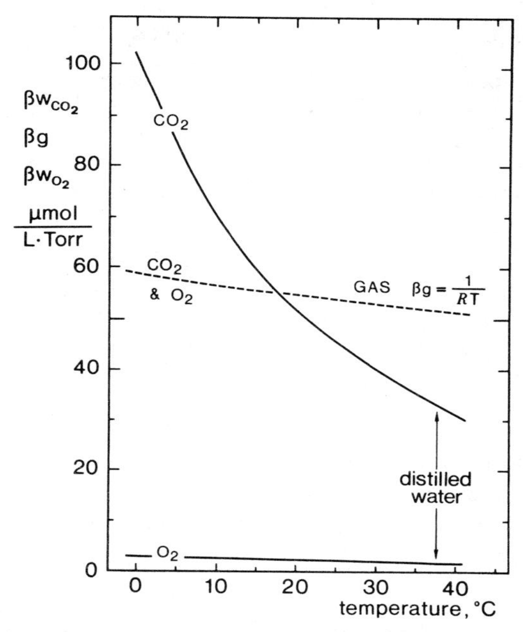

Fig. 3.3. Oxygen and carbon dioxide capacitance coefficients of distilled water and gas as functions of temperature. At 17.5 °C the CO_2 capacitance coefficients of distilled water and air are identical (from Dejours, 1981).

and the Congo eel, Heisler *et al.* (1982) have reported the following values: $P_{O_2} < 5$ Torr, $P_{CO_2} = 62$ Torr, pH = 5.6. Oxygenation is also very low in freezing water where a cover of ice prevents any exchange of O_2 with the atmosphere (Gnaiger and Forstner, 1983, pp. 256–259).

In sea water, carbon dioxide and oxygen concentrations and pressures, and pH, change with depth. These changes are illustrated by fig. 3.4 which shows in particular that the O_2 concentration falls with depth to almost zero at about 300–400 m and rises progressively at greater depths.

The penetration of light is extremely variable. First, it depends on the wave length; red light is quickly absorbed in the first meters beneath the surface, whereas blue-green light may penetrate to more than 50 m. Second, the depth of penetration depends on the turbidity of the water due to various mineral and organic particles and to microorganisms (Harvey, 1957, pp. 82–89). The light energy available entails photosynthesis which provokes circadian variations of P_{O_2} and C_{O_2}, of P_{CO_2} and C_{CO_2}, and of acid–base balance. During the day P_{O_2} and pH rise, and P_{CO_2} decreases; at night where animals and plants are only respiring, the

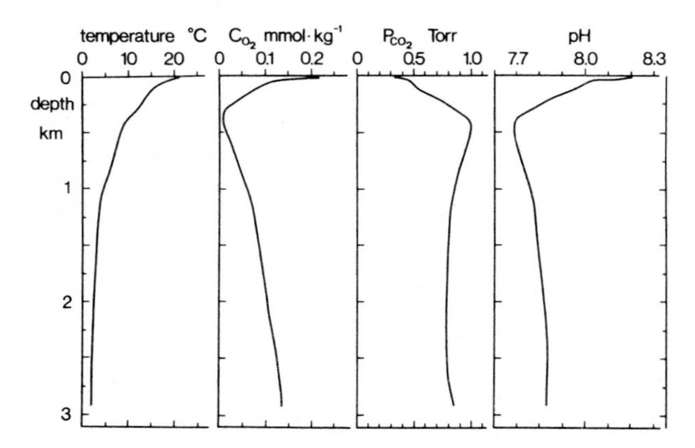

Fig. 3.4. Temperature, O_2 concentration, CO_2 partial pressure and pH of water in the eastern Pacific Ocean (4°00 S-82°00 W) at different depths. Data from Culberson and Pytkowicz (1970). The pH is recalculated at the actual local temperature and hydrostatic pressure; P_{CO_2} is calculated assuming a constant salinity of 34.8‰ (chlorinity = 19.3‰).

reverse is observed. Truchot and Duhamel-Jouve (1980) have studied these circadian changes; they have found a range of 10^{-3} to 2.7 Torr for CO_2; 10.2 to 7.3 for pH; 440 to 1.6 Torr for P_{O_2} in the water of a rockpool of the marine intertidal zone. Circadian variations also exist in the superficial layers of the open sea (Riley and Skirrow, 1975, pp. 144–147), but they are less than in rockpools or ponds.

Submarine thermal springs, also called deep-sea hydrothermal vents, have been discovered on the Galapagos Rift and on the East Pacific Rise. They are remarkable in two ways. (1) Their physicochemical conditions are extreme, since the water emerging from the fissures is hot, anoxic, acidic and contains an enormous amount of H_2S. This water mixes with normal surrounding oceanic water which is cold, 2 °C, contains oxygen at pressure of more than 50 Torr, no H_2S and has a pH value around 7.5; consequently, over a short distance, the physicochemical conditions change steeply. (2) In these special waters, a dense biomass is supported by the primary production of sulfide oxidizing bacteria. Large pogonophoran tube worms, clams, mussels and brachyuran crabs are found; some species of these groups have never been observed elsewhere. Their physiology is beginning to be known (see Arp and Childress, 1981; Spiess *et al.*, 1980; Hand and Somero, 1983; Hochachka and Somero, 1984; Childress *et al.*, 1987).

Atmosphere

As mentioned beforehand, the air, or any gas mixture, is a simpler physicochemical system than the waters. Equation (3.7) has shown that the capacitance of a gas phase is equal to $1/RT$, whatever the gas species or the barometric pressure. The only information one needs in order to calculate the concentration of a gas in a gas phase (eq. 3.6) is the temperature and barometric pressure, since

$$P_x = F_x \cdot (P_B - P^{T'}_{H_2O}) \tag{3.9}$$

where F_x is the mole fraction of a gas, P_B the barometric pressure, and $P^{T'}_{H_2O}$ is the saturating water vapor pressure at the body temperature T′ which prevails in the lung.

Actually, for a respiration physiologist, air is a very simple medium to deal with compared to waters with their countless variations of O_2 and CO_2, ionic composition and temperature.

The composition of open air is the same all over the world (F_{O_2} = 0.2095; F_{CO_2} = 0.00034). The ambient temperature and the relative humidity have to be taken into account if one has to deal with water loss in the process of external respiration. In our industrial world, however, the CO_2 mole fraction tends to rise; about 0.00034 in 1982 with small circannual variations (Revelle, 1982), it was 0.00030 at the beginning of this century. In industrial towns, the air is polluted, but its O_2 and CO_2 concentrations differ little from those of unpolluted air.

However, there are life-supporting atmospheres which differ markedly from open air. In caves where thousands or millions of bats live, the NH_3 concentration can be very high. In Guano Bat Cave studied by Mitchell (1964), the highest F_{NH_3} observed was 0.0018, that is about 1.4 Torr. At this partial pressure man survives only a few minutes, whereas it is harmless to the guano bats (Studier *et al.*, 1967). Possibly alveolar ventilation is low in these animals and alveolar and arterial P_{CO_2} are high; in turn hypercapnia may help to decrease the amount of free ammonia in the body fluids (Studier and Frasquez, 1969). Another microclimate is that found in the pouch of marsupials. In Virginia opossums' pouch with young, Farber and Tenney (1971) found F_{O_2} to average 0.142 and F_{CO_2} 0.043.

The O_2 and CO_2 properties of these media are summarized in table 3.1.

TABLE 3.1

Summary of the main physical differences between water and air for oxygen and carbon dioxide at 20 °C.

	Water	Air	Air/water
Diffusion coefficient ($cm^2 \cdot sec^{-1}$)			
D_{O_2}	0.000 025	0.198 (at 1 atm)	≃ 8000
D_{CO_2}	0.000 018	0.155	≃ 9000
Capacitance coefficient ($nmol \cdot ml^{-1} \cdot Torr^{-1}$)			
	1.82	54.7	≃ 30*
	51.4	54.7	≃ 1*
	40 500	54.7	≃ 1/750*
Krogh's constant ($nmol \cdot cm^{-1} \cdot sec^{-1} \cdot Torr^{-1}$)			
K_{O_2}	0.000 046	10.9 (at 1 atm)	≃ 2 × 10^5
K_{CO_2}	0.000 93	8.5	≃ 9000

* This ratio is the partition coefficient.

However, all their other properties differ quantitatively, often by one to three orders of magnitude (table 9.1, p. 118). These properties bear consequences for various bodily functions, such as nature of the nitrogenous end product, circulation, skeletal development, and the heat budget of aquatic and terrestrial animals (Chapter 9).

Soil environments

In the burrows, galleries, dens and caves, where various mammals, mainly rodents (see Arieli, 1979) and bats, live, as well as birds (see Boggs *et al.*, 1984), reptiles (see Johansen *et al.*, 1980) and amphibians (see Wood *et al.*, 1975), the atmospheres are asphyxic. Arieli (1979) has reported the action of soil texture, rain and temperature on burrow gas composition, and values of F_{O_2} and F_{CO_2} respectively below and above 0.10 have been recorded.

To a certain extent, some soils can be considered as porous bodies, hence the word porosphere (Vannier, 1983, p. 311). We may consider the porosphere as a collection of gas pockets communicating more or less freely with open air. Three factors govern the O_2 and CO_2 changes which may occur in these gas pockets: (1) the respiratory intensity of the various animal species and plant parts, like roots; (2) the degree of humidity which to a certain extent conditions whether the channels are free to open air; they may eventually be completely flooded; and (3) the chemical composition of the soil, which may be either acid (podzol) and consequently unable to fix CO_2 chemically, or calcareous (clay soil) and able to buffer some carbon dioxide. According to the relative importance of the gas and water phases and of the buffering properties of the soil, the simultaneous changes of P_{O_2} and P_{CO_2} due to respiration in the underground pockets are variable (Verdier, 1975; Vannier, 1983).

Figure 3.5 shows how the ambient P_{CO_2} and P_{O_2} values for animals living in various underground media change simultaneously. Let us suppose that animals are placed in a medium of composition I, at 20 °C, $PI_{O_2} = 150$ Torr, $PI_{CO_2} = 0.25$ Torr, the O_2 and CO_2 pressures prevailing at sea level in water-saturated air. This medium may be a gas pocket in an impermeable soil (curve 1). It may be a water pocket containing non-buffering water, as an acid water (curve 2). Curve 3 concerns a carbonated water; in this example the carbonate alkalinity is assumed to be 5 meq · L^{-1}. Curves 4 and 5 are for pockets containing both gas and water; but medium 4

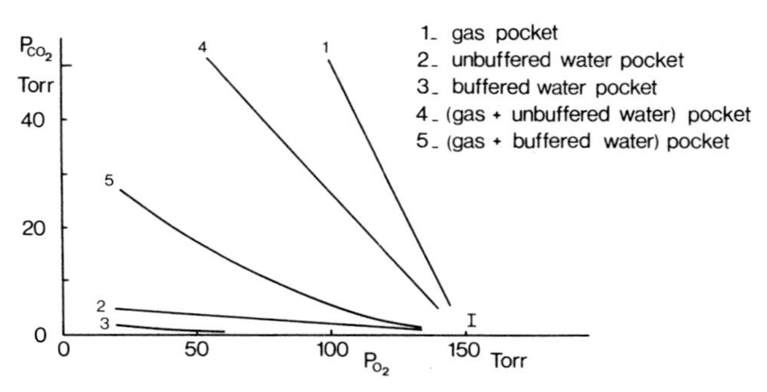

Fig. 3.5. P_{CO_2} *vs* P_{O_2} diagram. This diagram shows the simultaneous changes of P_{CO_2} and P_{O_2} of the milieu in pockets (closed systems) due to the respiration of small animals, those whose size is negligible compared to the size of the pocket. At the beginning the milieus have the P_{O_2} and P_{CO_2} values of water-saturated normal air at sea level. The respiratory quotient is assumed to be 1 and the temperature 20 °C. (From Rahn, 1966; Verdier, 1975; Dejours, 1981; Vannier, 1983.)

contains acid water as found in podzol, and medium 5 buffering water (carbonate alkalinity = 5 meq · L^{-1}) as it may be found in calcareous clay soil.

Now let us suppose that starting from point I, small animals, whose volume is negligible relative to the volume of medium available, use oxygen and produce carbon dioxide. The change of P_{CO_2} and P_{O_2} as a function of time will depend on the nature of the medium, and precisely on its capacitance (a respiratory quotient of one is assumed). The figure shows that in the gas phase, medium 1, P_{CO_2} builds up as much as P_{O_2} falls; P_{CO_2} increases much less in medium 2 (acid water) and still less in carbonated water (medium 3). For animals such as collemboles which may live in a gas pocket bathing either in an acid water (medium 4, a podzolic soil) or in a carbonated water (medium 5, *e.g.* a calcareous clay soil), the change of P_{CO_2} *vs* P_{O_2} is *intermediary* between those observed in air-breathers (medium 1) and water-breathers (medium 2 and 3). The increase of P_{CO_2} in medium 5, a gas pocket in a carbonated water with a high capacitance for CO_2 is lower than in medium 4, since the buffered water represents a sink for the carbon dioxide produced. The situation becomes more complex if there are connections between the soil and open air. To date, relatively few studies have been conducted on the gas exchange characteristics of the edaphic animals.

In the edaphic environment, the properties for heat dissipation also differ. This is very important, for example, for the development of a clutch of reptilian eggs layed at a certain depth in the soil or sand.

CHAPTER 4

A typical air breather, the human, *versus* a typical water breather, the dogfish. Dual breathers

Summary

The respiratory traits of air breathers are different from those of water breathers. Comparing them is made difficult by certain factors which impose their own respiratory constraints: (1) *Temperature*. Changes modify the energy metabolism, the oxygen consumption and the acid–base balance (ABB) of the body fluids. (2) *Size*. In a given zoological group, the higher the body mass, the lower the mass-specific energy metabolism. (3) *Zoological group*. Two animals of the same size, at the same temperature, at the same stage of development, may have quite different rates of energy metabolism.

These factors can be circumvented. Taking oxygen consumption as a reference for aerobic animals, ventilation and blood flow for water- and air breathers (as long as the external O_2 consumption takes place only via the ventilatory process) can be normalized by dividing ventilation and blood flow by the O_2 consumption to give an $\dot{M}_{O_2}$-specific ventilation and an $\dot{M}_{O_2}$-specific cardiac output, $\dot{V}w \cdot \dot{M}_{O_2}^{-1}$, $\dot{V}G \cdot \dot{M}_{O_2}^{-1}$ and $\dot{V}b \cdot \dot{M}_{O_2}^{-1}$, respectively, the specific ventilation and specific blood flow.

The ABB characteristics must be measured at the same temperature; if not, the blood characteristics resulting from a temperature difference can be calculated for an arbitrary temperature. A temperature-corrected ABB may be different from the real ABB observed at the arbitrary temperature.

The main differences between water- and air breathers are:

(1) The specific ventilation is lower in air breathers than in water breathers, because, for a given O_2 tension, the O_2 concentration is much higher in air than in water.

(2) The value of P_{CO_2} is higher in air breathers than in water breathers.

(3) The P_{CO_2} value for dual breathers is intermediary. The more an animal is a pulmonary breather, the higher is its P_{CO_2} value.

(4) There is no systematic difference of the pH value between water-, dual and air breathers. Since P_{CO_2} values vary among mammals, the relative constancy of pH implies variations in the concentration of body fluid carbonates.

Choice of models

If a 'typical' air breather is to be compared to a 'typical' water breather, what is wanted are two animals differing only in the ambient medium breathed, but alike in other ways: body mass, temperature, activity level, intensity of energy metabolism. A tortoise and a dogfish would be an ideal pair. Unfortunately complete data on O_2- and CO_2 net fluxes, flow rates of air and blood, O_2- and CO_2 tensions and concentrations in arterialized and mixed venous blood are available for very few species.

Here the exclusive water breather is the dogfish. In *Scyliorhinus stellaris*, studied by Baumgarten-Schumann and Piiper (1968), 95% of the external O_2 and CO_2 exchanges take place in the gills, and the rest transcutaneously. Analogous values of the cutaneous respiration of *S. canicula*, a closely related species, have been reported in Toulmond *et al.* (1982)*.

The exclusive air breather is the one most studied, the human, and the description is based on the data of Farhi and Rahn (1960) completed by values drawn from other reports. The comparison with the dogfish is made difficult by the fact that the intensity of energy metabolism is much higher in the human for reasons related to size, body temperature, and zoological relation (a homeotherm *vs* a poikilotherm). The use of normalized terms such as the $\dot{M}_{O_2}$-specific flow rates, $\dot{V}w \cdot \dot{M}_{O_2}^{-1}$, $\dot{V}air \cdot \dot{M}_{O_2}^{-1}$, $\dot{V}b \cdot \dot{M}_{O_2}^{-1}$, permits the comparison. A value of $\dot{V}w \cdot \dot{M}_{O_2}^{-1}$, for example, gives the volume of water which must be pumped through the gills to deliver one unit amount of oxygen to the body. These ratios $\dot{V}w \cdot \dot{M}_{O_2}^{-1}$, $\dot{V}_G \cdot \dot{M}_{O_2}^{-1}$, $\dot{V}b \cdot \dot{M}_{O_2}^{-1}$ for water, gas and blood can be called briefly specific ventilation and specific blood flow. Other terms still often used are the water- or air convection requirements and the blood flow requirement, or the ventilatory and circulatory equivalents (Dejours *et al.*, 1970). On the other hand, the intensive variables such as partial pressures and the dimensionless extraction coefficients for the two animals can be compared directly.

The discussion is based on the data of table 4.1, for animals in steady

---

* Actually, the $\dot{M}_{O_2}$-specific flow rates are valid only in animals in which all the O_2 uptake takes place in gills or lungs and not transcutaneously (see Dejours, 1973). In some fishes, the cutaneous O_2- and CO_2 transfers are not negligible (see Nonnotte and Kirsch, 1978; Feder and Burggren, 1985).

state, at rest. The values for each species are coherent, although not all are certain; the comparison between the animals can nonetheless be made confidently, because the differences between them are so large. Finally, table 4.1 may appear cumbersome, but it could have been longer: only the data demonstrating the main differences or similarities between a typical air breather and a typical water breather are given.

Comparison of the two models: dogfish and human

(1) The partial pressures of oxygen in the inspired water and air, $P_{I_{O_2}}$, are the same, while the corresponding O_2 concentration, $C_{I_{O_2}}$, is much higher in air than in water (table 4.1, lines j and m).

Since the concentration of available O_2 is much higher in air than in water, the specific ventilation (line h) can be and actually is much lower in the air breather than in the water breather, although the extraction coefficient of O_2 from the medium (line q) is lower in the air breather than in the water breather. These differences are general between all the water-breathing and air-breathing species of the various classes and phyla which have been studied (Dejours *et al.*, 1970).

(2) The CO_2 solubility in water and its capacitance in air (line h′) are similar. However, in water the effective CO_2 solubility or capacitance (line g′), is often higher than the physical solubility (line h′), since the exact value of the capacitance depends on the status of the buffer system in the water (see p. 21). It follows from these similar capabilities of air and water to fix CO_2, and from the fact that the specific ventilation is lower in the air- than the water breather, that the difference of P_{CO_2} between expired and inspired media is several times higher in air breathers than in water breathers. There is no exception to this law, as long as one considers exclusively pulmonate air breathers and exclusive water breathers.

For various reasons, P_{O_2} and P_{CO_2} in the postbranchial and post-pulmonary blood are not very different from their values in the gill water or in the alveolar gas, although the mechanisms of the O_2 and CO_2 exchanges in gills and in lungs are very complex (see, for example, Piiper and Scheid, 1975). Figure 4.1, whose values are taken from table 4.1, shows that the P_{CO_2} values in the various compartments of the whole body are much higher in the air breather than in the water breather. If the animals breathed a hypercapnic milieu, then the whole set of points I, E, a

TABLE 4.1

Respiratory variables in a typical water breather, a dogfish, and a typical air breather, a human. For abbreviations see pp. xiii–xvi. Main data from Baumgarten-Schumann and Piiper (1968) for the dogfish, and from Farhi and Rahn (1960) for the human. Some figures were estimated in order to have a more complete and comparable list of data. For Pa_{CO_2} and pHa, the values in parentheses, recalculated for an *in vivo* decrease of temperature from 38 to 16 °C, may be compared to the values observed in the dogfish.

			Dogfish	Human	Human/ dogfish
a	Body temperature	°C	16	37.0	
b	Body mass	kg	2.18	70.0	
c	$\dot{M}_{O_2}$	$mmol \cdot min^{-1}$	0.0623	12.06	
d	$\dot{M}_{CO_2}$	$mmol \cdot min^{-1}$	0.0602	10.25	
e	$\dot{V}w$ and $\dot{V}g$	$L \cdot min^{-1}$	0.425	7.40	
f	$\dot{V}b$	$L \cdot min^{-1}$	0.048	5.20	
g	$\dot{V}w \cdot \dot{V}b^{-1}$ and $\dot{V}g \cdot \dot{V}b^{-1}$		8.85	1.42	
h	$\dot{V}w \cdot \dot{M}_{O_2}^{-1}$ and $\dot{V}g \cdot \dot{M}_{O_2}^{-1}$	$L \cdot mmol^{-1}$	6.82	0.61	0.09
i	$\dot{V}b \cdot M_{O_2}^{-1}$	$L \cdot mmol^{-1}$	0.770	0.43	0.56
j	$P_{I_{O_2}}$	Torr	149	149	
k	$P_{E_{O_2}}$	Torr	56	119	
l	$\Delta P_{I, E_{O_2}}$	Torr	93	30	
m	$C_{I_{O_2}}$	$mmol \cdot L^{-1}$	0.237	7.70	32.5
n	$C_{E_{O_2}}$	$mmol \cdot L^{-1}$	0.089	6.15	
o	$\Delta C_{I, E_{O_2}}$	$mmol \cdot L^{-1}$	0.148	1.55	
p	αw_{O_2} and βg_{O_2}	$mmol \cdot L^{-1} \cdot Torr^{-1}$	0.00159	0.0517	32.5
q	Ew_{O_2} and Eg_{O_2}		0.624	0.201	0.32
r	$P_{A_{O_2}}$	Torr		105.5	
s	Pa_{O_2}	Torr	49	93.0	
t	$P\bar{v}_{O_2}$	Torr	10	38.0	
u	$\Delta P\bar{a}, \bar{v}_{O_2}$	Torr	39	55.0	
v	Ca_{O_2}	$mmol \cdot L^{-1}$	1.83	8.71	4.8
w	$C\bar{v}_{O_2}$	$mmol \cdot L^{-1}$	0.491	6.39	
x	$\Delta Ca, \bar{v}_{O_2}$	$mmol \cdot L^{-1}$	1.34	2.32	
y	$\beta a, \bar{v}_{O_2}$	$mmol \cdot L^{-1} \cdot Torr^{-1}$	0.0343	0.0422	
z	Eb_{O_2}		0.732	0.266	

a′	PI_{CO_2}	Torr	0.7	0.25	
b′	PE_{CO_2}	Torr	1.3	26.33	
c′	$\Delta PE, I_{O_2}$	Torr	0.6	26.08	43.3
d′	CI_{CO_2}	$mmol \cdot L^{-1}$	2.337	0.015	
e′	CE_{CO_2}	$mmol \cdot L^{-1}$	2.480	1.361	
f′	$\Delta CE, I_{CO_2}$	$mmol \cdot L^{-1}$	0.143	1.344	
g′	$\beta E, I_{CO_2}$	$mmol \cdot L^{-1} \cdot Torr^{-1}$	0.238	0.0517	0.22
h′	αw_{CO_2} and βg_{CO_2}		0.049	0.0517	1.0
i′	PA_{CO_2}	Torr		38.0	
j′	Pa_{CO_2}	Torr	2.0	38.0	19
k′	$P\bar{v}_{CO_2}$	Torr	2.6	44.0	
l′	$\Delta P\bar{v}, I_{CO_2}$		1.9	44.0	27
m′	$\Delta Pv, a_{O_2}$	Torr	0.6	6.0	
n′	Ca_{CO_2}	$mmol \cdot L^{-1}$	3.08	21.34	
o′	$C\bar{v}_{CO_2}$	$mmol \cdot L^{-1}$	4.37	23.31	
p′	$\Delta C\bar{v}, a_{CO_2}$	$mmol \cdot L^{-1}$	1.29	1.97	
q′	$\beta\bar{v}, a_{CO_2}$	$mmol \cdot L^{-1} \cdot Torr^{-1}$	2.16	0.328	
r′	pHa		7.78 (7.44*)	7.40	
s′	pHa-pN		0.62	0.59	

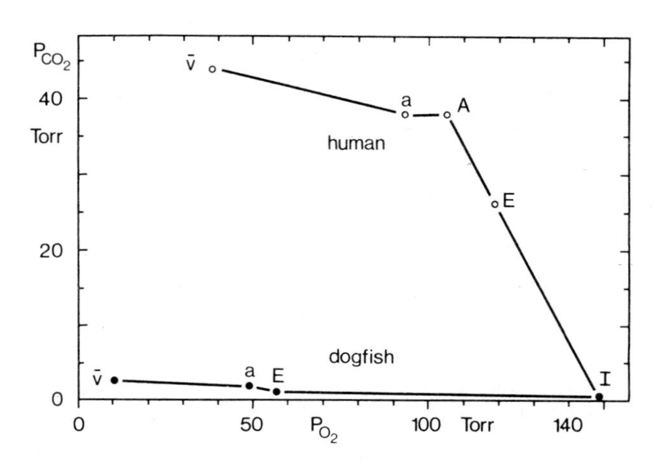

Fig. 4.1. P_{CO_2} *vs* P_{O_2} diagram. Symbol I for inspired water and gas (actually the inspired water and air compositions are slightly different, but this cannot be shown graphically); E, expired water and air; A, alveolar gas; a, arterial blood; v̄, mixed venous blood. Data in human and dogfish taken from table 4.1.

and $\bar{v}$ would be shifted upwards; yet the difference between $P\bar{v}_{CO_2}$ and PI_{CO_2} ($\Delta P\bar{v}$, I_{CO_2}, line l′) would remain much larger in the air breather than in the water breather. This also is a general law.

The respective positions of points I, E, A, a and $\bar{v}$ of the air breather and of I, E, a and $\bar{v}$ of the water breather, which may vary among species, or from one animal to another of the same species with the experimental conditions, could be commented on at length, but such a discussion is not within the scope of this book.

(3) In spite of the enormous difference between the arterial P_{CO_2} values in the human and the dogfish, their acid–base balance is not fundamentally different. Table 4.1 does not apparently support this statement, since the pH values are 7.40 in the human and 7.78 in the dogfish. In fact these values cannot be directly compared, because of the very marked effect of thermal change on acid–base balance. The pH of neutral water, pN, is temperature-dependent. At the 37 °C temperature of the human, pN = 6.81; at the 16 °C temperature of the dogfish, pN = 7.16. Thus pH of neutral water, pN, increases by 0.35 pH unit when the temperature decreases by 21 °C, that is about 0.016 pH unit per °C. Consequently the blood pH values of the human and the dogfish at their respective body temperature exceed the pH of neutral water, pH − pN, by 0.62 and 0.59 pH unit, very close values.

Other pH values have been recorded in other animals; the blood pH is consistently on the alkaline side, and for most species this positive pH − pN difference is more or less constant over the whole body temperature range. It is why Rahn and his coworkers introduced the concept of *constant relative alkalinity* (Howell *et al.*, 1970). The intracellular pH is also temperature-dependent with the same slope as the extracellular pH; but systematically, the intracellular pH is lower than the extracellular pH and is, actually, close to the pH of neutral water, pN (Rahn *et al.*, 1975).

The pattern of the pH temperature-dependency is a very important phenomenon because it ensures a constant dissociation of proteins and the stability of the enzymatic activity (Rahn and Reeves, 1982; Somero and White, 1985). The reasons for this pH *vs* temperature behavior are complex. First of all, *in vitro* blood and intracellular fluid follow Rahn's rule, mainly because of dissociation of the alpha imidazole group of histidine as a function of temperature (Reeves, 1972, 1977). However, organisms are open systems and the resemblance to an *in vitro* system is intriguing. Details about the ways water-breathing and air-breathing animals follow Rahn's rule can be found in Reeves (1977)*.

Temperature variations change P_{CO_2} values as well as pH values. Thus, again, the blood P_{CO_2} of two animals cannot be directly compared when their body temperature is different. Tables 4.1 and 4.2 show what happens to pH and P_{CO_2} values of *in vitro* human blood cooled to 16 °C. It may be seen that P_{CO_2} is 12.5 Torr whereas it is 2.0 in the dogfish; P_{CO_2} is systematically higher in air breathers than in water breathers (see the following section).

(4) The first three comments are the most important since they are general to all air breathers compared with all water breathers. We will now compare some heart and blood data in humans and dogfish, but the differences observed cannot be generalized. The main fact is that the blood O_2 concentration is higher in man than in the dogfish; this is by no means a characteristic of air- *vs* water breathers: some fish, *e.g. Tuna*, and some annelids, *e.g. Arenicola*, may have very high blood O_2 concentrations. Because in man, Ca_{O_2} is 4.8 times higher than in the dogfish (Table 4.1, line v), his oxygen-specific blood flow and arteriovenous O_2 extraction are lower. That the specific blood flow and the O_2 extraction from blood depend on the O_2 concentration in the arterial blood is a general law (Dejours *et al.*, 1970).

From our examples, man and dogfish, it follows that the ratios ventilatory flow/blood flow ($\dot{V} \cdot \dot{V}b^{-1}$), $\dot{V}g \cdot \dot{V}b^{-1}$ for the air breather and $\dot{V}w \cdot \dot{V}b^{-1}$ for the water breather are respectively 1.42 and 8.85 (line g). This discrepancy is not a general law because $\dot{V}g \cdot \dot{M}_{O_2}^{-1}$ and $\dot{V}b \cdot \dot{M}_{O_2}^{-1}$ are extremely variable as a function of the values of CI_{O_2} and Ca_{O_2} (figs. 5.4 and 5.5). This ventilation/perfusion ratio is a very important factor in the gas exchange in the lung (Rahn, 1949; Farhi, 1987) as well as in the gill (Piiper and Scheid, 1975). For the lung, the $\dot{V}A \cdot Vb^{-1}$ ratio determines the alveolar gas composition (see Fenn *et al.*, 1946; Rahn and Fenn, 1955). For the gill, the ratio $\dot{V}w \cdot \dot{V}b^{-1}$ is also very important in branchial

* There seem to be exceptions to Rahn's rule for the extracellular fluids of some aquatic animals (Reeves, 1977; Dejours, 1981; Truchot, 1987b) and of some reptiles, as *Varanus exanthematicus* (Wood *et al.*, 1981). But the intracellular pH of these animals may have a normal temperature dependency (Heisler, 1982; Gaillard and Malan, 1983; Rodeau, 1984). For appraisals of the changes of ABB in body fluids, see Heisler (1986) and Truchot (1987b).

TABLE 4.2

Some respiratory variables in selected water breathers and air breathers. The 37 °C human blood values are taken from table 4.1. The values of P_{CO_2} and pH of human blood at 25, 16 and 8 °C were calculated according to Malan (1977) for blood *in vitro* without gas phase.

1	2	3	4	5	6	7	8	9	10	11	12
Ref.		T (°C)	$P_{I_{O_2}}$ (Torr)	$P_{E_{O_2}}$ (Torr)	$\Delta P_{I,E_{O_2}}$ (Torr)	$\Delta P_{E,I_{CO_2}}$ (Torr)	Pb_{O_2} (Torr)	Pb_{CO_2} (Torr)	pHb	$[HCO_3^-]b$ (meq · L^{-1})	$\dot{V}/\dot{M}_{O_2}$ (L · mmol^{-1})
WATER BREATHERS											
11	Lugworm, *Arenicola marina*	22	159					1.7	7.29	1.55	
12	*Arenicola marina*	15	160	35	125	0.3					5.0
8	*Octopus dofleini*	11	127	94	33		78	3.1	7.13		17.3
13	Shore crab,	26						4.1	7.53	5.5	
	Carcinus maenas	16	156					3.0	7.66	6.6	
		10						2.1	7.75	6.5	
		5						1.7	7.90	7.8	
10	Dungeness crab, *Cancer magister*	8	141	97	44		75	1.7	7.9		14.2
1	Dogfish, *Scyliorhinus stellaris*	16	149	56	93	0.6	49	2.0	7.78		6.8
AIR BREATHERS											
14	Snail*,	25	152	137	15	11.5	137	11.1	7.86		
	Otala lactea	15	154	146	8	9.1	146	9.5	8.04		
		5	156	148	8	5.3	148	5.2	8.28		
	Otala, dormant	25	152	38	114	63			7.21		
5	Terrestrial crab, *Gecarcinus lateralis*	21	155					8.9	7.49	11.4	
6	Fiddler crab,	30	155					16.9	7.49	19.2	
	Uca pugilator	20						9.7	7.76	23.2	
	Uca pugilator	10						5.1	7.89	21.1	

3	Coconut crab, *Birgus latro*	29	151				27	9	7.70	12.7	2.8
9	Spider (tarantula), *Eurypelma calif.*	25	155				28	10.6	7.57		
7	Red-eared turtle,	30						31.9	7.56	33.0	0.51
	Pseudemys scripta el.	25						27.5	7.61		
	Pseudemys scripta el.	20	155					22.7	7.67	34.8	0.78
	Pseudemys scripta el.	10						14.1	7.76	33.1	1.39
2	Pekin duck	41	144	10.8	36	26	91	33	7.48	23	0.54
4	Human	37	149	119	30	26.1	93	38	7.40		0.59
	Human blood *in vitro*	25						23.5	7.57		
	Human blood *in vitro*	16						12.5	7.74		
	Human blood *in vitro*	8						7.4	7.89		
15	Hamster, *Cr. cricetus*	37						45.3	7.40		
	Hamster, hibernating	9						36.1	7.57		
	Hamster, *in vitro*	9						10.1	7.84		

Columns 4–7 concern external breathing. Columns 8–11 concern blood; it may be arterial blood as in the dogfish, the snail, the duck and the human; in the turtle the blood drawn from the subclavian artery is actually a mixture of arterialized and venous blood; for *Arenicola* it is venous blood; for the crustaceans it may be pre- or post-branchial blood. All blood values have the subscript b without specification. Note the wide range of temperature; thus the Pb_{CO_2} and the ABB values are not directly comparable.

Data for the dogfish and the human (37 °C) drawn from table 4.1.

1 Baumgarten-Schumann and Piiper (1968)
2 Bouverot *et al.* (1979)
3 Burggren and McMahon (1981)
4 Farhi and Rahn (1960)
5,6 Howell *et al.* (1973)
7 Jackson *et al.* (1974)
8 Johansen and Lenfant (1966)
9 Loewe and Brauer de Heggert (1979)
10 McMahon *et al.* (1979)
11 Toulmond (1977)
12 Toulmond and Tchernigovtzeff (1984)
13 Truchot (1973)
14 Barnhart (1986)
15 Malan *et al.* (1973); Malan (1982)

For the red-eared turtle the data at 25 °C are interpolated for table 4.3 and fig. 4.3.

* The snail *Otala* exchanges O_2 and CO_2 by diffusion. Column 5 refers to pulmonary gas; $\Delta P_{E,I}$ (columns 6 and 7) refers to the difference between air and pulmonary gas.

respiratory exchanges, but no simple relation can be given because of the disposition of the gills and of the nature of the water and blood flows which can stream in opposite directions (countercurrent disposition).

Are other water and air breathers comparable to these models?

Among water breathers there are some differences in their O_2- and CO_2 status, related to the actual ambient P_{CO_2}, to the CO_2-buffering properties of the water they live in and to the anatomical disposition of their respiratory system. However, as mentioned above, blood P_{CO_2} is never very high (table 4.2) or, more exactly, never very different from the inspired P_{CO_2}, because the water capacitance is much higher for CO_2 than for O_2. A large change of P_{O_2} between inspired and expired waters can lead to only a few torrs increase of CO_2 (Rahn, 1966). There are, however, two circumstances where P_{CO_2} may be high in aquatic animals: (1) when they live in a water with a high P_{CO_2} value, as in certain aquatic biotopes (p. 25); (2) when they live in a hyperoxic water, because hyperoxia depresses breathing and thus increases the inspired-to-expired P_{O_2} and P_{CO_2} differences, $\Delta P_{I,E_{O_2}}$ and $\Delta P_{I,E_{CO_2}}$. In all cases, however, $\Delta P_{E,I_{CO_2}}$ is much smaller than $\Delta P_{I,E_{O_2}}$.

For air breathers, the situation is completely different, since the O_2- and CO_2 capacitances are similar. Thus the O_2- and CO_2 differences between inspired and expired gas are close and related only to the respiratory quotient. In our example of typical air breather, the human, the alveolar and arterial P_{CO_2} are about 38 Torr. This value is by no means a constant and varies with several factors: posture, sleep and awakeness, sexual cycle, diet, exercise, altitude, *etc.* In other species of mammals, birds and reptiles, the mean values of pulmonary and blood P_{CO_2} may be somewhat different from the human value, but the P_{CO_2} values are always more than 10 Torr, except in some special conditions, such as very high altitude or extreme heat, which will be taken up later. Obviously, birds and mammals breathe much less than the water breathers to get a given quantity of oxygen, because air contains much more oxygen than water does for the same O_2 tension.

Table 4.2 gives a selection of respiratory variables in air breathers. In the three examples of terrestrial crustaceans, blood P_{CO_2} is higher than in water breathers, but the values are lower than in air breathing vertebrates;

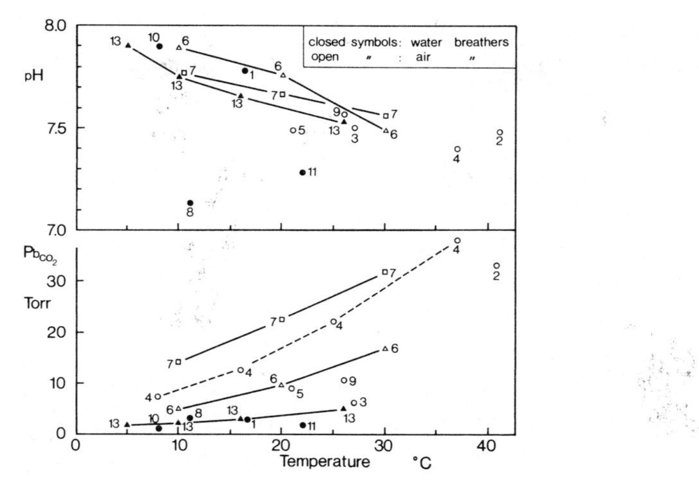

Fig. 4.2. Blood pH and P_{CO_2} as functions of body temperature in water breathers (closed symbols) and air breathers (open symbols). The numerals next to each point correspond to the reference number of table 4.2, from which the data are taken. This figure shows that Pb_{CO_2} is higher in air breathers than in water breathers (with the possible exception of the coconut crab), but there does not seem to be any systematic pH difference between the two groups. It also shows that pH decreases and Pb_{CO_2} increases when temperature rises. The Pb_{CO_2} points at 25, 16 and 8 °C connected by a dashed line concern human blood and are calculated from the value at 37 °C, that is 38 Torr, when blood is cooled in a closed system (Malan, 1977); this Pb_{CO_2} fall and the concomitant pH increase (not plotted) occur in blood which, leaving the heart at 37 °C, is cooled in the limbs exposed to a cold environment.

more examples are given by Cameron and Mecklenburg (1973), Innes and Taylor (1986) and Truchot (1987a). It should be noted, however, that the P_{CO_2} value increases with temperature. The same temperature-dependence is observed in the turtles *Pseudemys scripta* (Jackson *et al.*, 1974) and *Chrysemis picta* (Funk and Milsom, 1987) (fig. 4.2). Consequently it is compulsory to compare animals at the same temperature (Rahn and Garey, 1973), or, if that is not feasible, the complex action of temperature on respiratory variables must be taken into account.

Figure 4.3 shows how important the factor temperature is for the determination of pulmonary P_{CO_2} and P_{O_2} and for comparison with analogous variables in water breathers. In this figure, the curves for human and for dogfish are the same as in fig. 4.1. Some remarkable data obtained in the air-breathing snail *Otala lactea* have been added (Barnhart, 1986). This pulmonate exchanges O_2 and CO_2 by diffusion through the pneumostome which is the connection between the lung cavity and the ambient medium. Thus, in this animal, the point A corresponds to the pulmonary gas and no expired gas can be defined. Presumably some respiratory exchanges may also occur through the exposed tegument, and could, in part, explain why P_{CO_2} values are so low. But the most important point is that $P_{A_{O_2}}$ and $P_{A_{CO_2}}$ values vary with temperature as shown by table 4.2 and fig. 4.3. The lowest value of $P_{A_{CO_2}}$ is 5.3 Torr in the animal at 5 °C, a value which is not much higher than those observed in some water breathers. But their exists a striking difference between an air-breathing animal, here the snail *Otala*, and the water-breathing dogfish. The two animals can be compared because the snail and the dogfish were studied at about the same temperature, 16 and 15 °C respectively. We will compare the pulmonary gas of the snail to the midpoint between inspired (I) and expired water (E) of the dogfish, assuming that the composition of the midpoint between I and E is an estimate of an operative composition of the water around the branchial areas. Table 4.3 shows that the fall of P_{O_2} between ambient air and the lung, $\Delta P_{I,\ A_{O_2}}$, in the snail is much lower than the fall between ambient water and the branchial water in the dogfish, $\Delta P_{I,\ br_{O_2}}$: 8.5 Torr and 46.5 Torr respectively. For CO_2, the corresponding changes ΔP_{CO_2} between the ambient milieu and the gas exchange area are 8.8 in the snail and 0.6 Torr in the dogfish. A general rule is: the change of P_{O_2} and P_{CO_2}, ΔP_{O_2} and

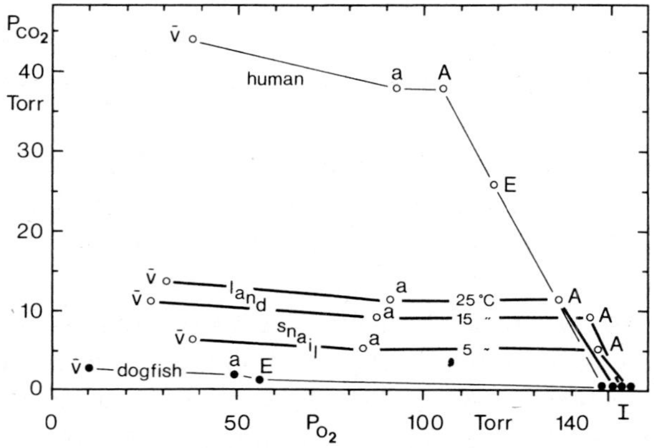

Fig. 4.3. P_{CO_2} *vs* P_{O_2} diagram. See legend of fig. 4.1 and table 4.1 for the human and the dogfish. The curves marked 5, 15 and 25 °C concern the land snail, *Otala lactea*. The symbols A for this pulmonate which exchanges O_2 and CO_2 by diffusion concern the pulmonary gas (see text and table 4.3 for details).

TABLE 4.3

P_{O_2} and P_{CO_2} values in inspired milieus, in expired and alveolar gas and in arterial blood of the dogfish at 16 °C and in lung gas A of the land snail *Otala* at 15 °C. Pbr_{O_2} designates the O_2 pressure at the gill surface, which is taken as the mid value between inspired and expired water's P_{O_2} values. This estimation has not been made for CO_2, because the difference is not important and because the relation between C_{CO_2} and P_{CO_2} is not linear. Data from Baumgarten-Schumann and Piiper (1968) for the dogfish, and from Barnhart (1986) for the snail *Otala*. See text for interpretation.

		Dogfish *Scyliorhinus*	Snail *Otala*
Temperature	(°C)	16	15
PI_{O_2}	(Torr)	149	154
PE_{O_2}	(Torr)	56	
Branchial P_{O_2}, Pbr_{O_2}	(Torr)	102.5	
PA_{O_2}	(Torr)		145.5
$\Delta PI, br_{O_2}$	(Torr)	46.5	
$\Delta PI, A_{O_2}$	(Torr)		8.5
PI_{CO_2}	(Torr)	0.7	0.3
PE_{CO_2}	(Torr)	1.3	
PA_{CO_2}	(Torr)		9.1
$\Delta PE, I_{CO_2}$	(Torr)	0.6	
$\Delta PA, I_{CO_2}$	(Torr)		8.8
Pa_{CO_2}	(Torr)	2.0	9.2
Pa, I_{CO_2}	(Torr)	1.3	8.9

ΔP_{CO_2}, between the ambient milieus and the pulmonary or branchial exchanging surface areas are of the same order of magnitude in air breathers, whereas ΔP_{O_2} is far less than ΔP_{CO_2} in water breathers.

The conclusion is that to differentiate air breathers from water breathers, one has to consider *simultaneously* the changes undergone by P_{CO_2} and P_{O_2} between the ambient medium and the pulmonary or branchial medium.

If the pulmonary or branchial medium is inaccessible and its composition cannot be estimated, then one can compare the variations of P_{CO_2} between the ambient milieu and the blood; in our example (table 4.3), 1.3 Torr for the dogfish; 8.9 Torr for the snail. This comparison of blood to water P_{CO_2} is possible because there is no marked difference between the pulmonary or branchial operative P_{CO_2} and arterialized blood P_{CO_2}, whereas as shown by fig. 4.3 the difference for P_{O_2} is enormous, mainly because of the diffusion limitation in the case of the air-breathing snail and the water-breathing dogfish.

Dual breathers

Animals display a variety of respiratory surfaces and body plans (Tenney, 1979). The respiratory surfaces may be reduced to the simple contact between a single-celled animal or the outer cells of small metazoans with the surrounding milieu: water, water-saturated air, rarely more or less dry air. In many phyla the tegument is the respiratory surface. Among specialized respiratory surfaces are gills; diverticulae of the digestive tract; particular development of the skin; openings to the outside, as the spiracles of insects, from which tracheae spread into the body; lungs, of which there are many variations: sac-like lungs in amphibians and reptiles, alveolar lungs in mammals, tubular or parabronchial lungs in birds. The avenue of respiratory exchanges is referred to as a modality. Animals may be unimodal, bimodal, trimodal; for example the urodele *Siren lacertina* is trimodal, for respiration is cutaneous, branchial and pulmonary. There are many combinations of modalities when one takes into account specialized exchange surfaces.

However, the main factor which dominates the respiratory physiology of an animal is not so much the modality of the external respiration as the nature of the milieu in which it lives. Whatever the anatomical features of their respiratory surfaces, many animals have the physiological characteristics either of water breathers or of air breathers, as discussed in the previous section. But many animals in various groups exchange O_2 and CO_2 with water *and* with air, either alternately or simultaneously (Hughes, 1975). These are called dual breathers.

The lowest P_{CO_2} values in blood are observed in water breathers; the highest in air breathers (tables and figs. 4.2, 4.3). The difference between exclusive air breathers and the exclusive water breathers is quite clear. What about dual breathers? Their blood acid–base balances (ABB) cannot be directly compared since the temperature at which measurements in different animals are made is not usually the same and as said above, thermal changes markedly influence ABB characteristics. However, one can estimate the blood ABB characteristics at a given conventional temperature calculated according to the principles introduced by Reeves

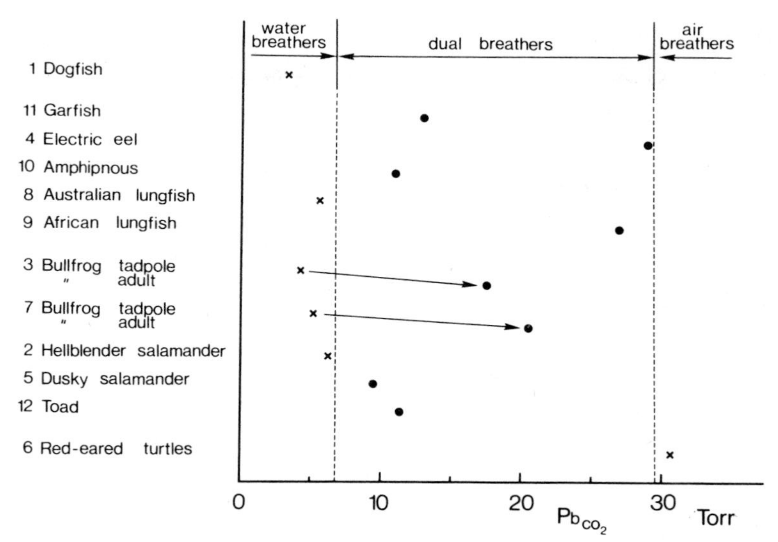

Fig. 4.4. Representation of Pb_{CO_2} at 25 °C of selected exclusive water- and air breathers and of dual breathers. The values are taken from table 4.4. (●) Dual breathers; (×) the exclusive or almost exclusive water breathers and the exclusive air breather (the turtle). The arrows show the change of Pb_{CO_2} which occurs in the metamorphosis of a water-breathing tadpole into a dual breathing adult bullfrog. The numbers before the animals' names are the reference numbers given in table 4.3.

(1976). The conventional temperature of 25 °C was chosen because several observations had been made at that temperature, but it is not certain that the ABB values calculated for 25 °C will be identical to those of the animal actually examined at 25 °C. Within a small range of thermal change, however, *in vivo* and *in vitro* ABB changes do not differ greatly (Truchot, 1987b).

Figure 4.4 shows the wide range of Pb_{CO_2} values in dual breathers. If the animal has a predominantly pulmonary ventilation, its P_{CO_2} value is high because it can breathe much less to be adequately oxygenated than a more aquatic breather. This is quite clear for the Australian lungfish, with a predominantly pulmonary respiration, and for the electric eel, whose buccal cavity is endowed with an extensively diverticulated and richly perfused mucosa which functions as an air lung (Garey and Rahn, 1970).

Table 4.4 and fig. 4.4 show two observations of what happens during the metamorphosis of a bullfrog tadpole, a strict water breather, into an adult bullfrog with some pulmonary respiration. When the gills disappear and the lungs develop and become functional, the animal, still living in

TABLE 4.4

Some respiratory variables in selected exclusive water breathers (dogfish, tadpoles), in almost exclusive water breathers (Australian lungfish, hellbender salamander), in various dual breathers, and in an exclusive air breather (red-eared turtle). The actual temperature of measurements is indicated in column 4; when the animals' temperature was different from 25 °C, P_{CO_2} and pH in the blood have been computed for 25 °C (in italics) according to Reeves (1976). No P_{CO_2} value in the inspired milieu is indicated. In air it is near 0.3 Torr, in water it is near 0.3 Torr if the water is very well equilibrated with air; unfortunately in many articles detailed information is not given about P_{CO_2} in the ambient water.

1	2	3	4	5	6	7	8	9	10
Ref.	Species	Respiratory surfaces	T (°C)	$P_{I_{O_2}}$ (Torr)	$P_{b_{O_2}}$ (Torr)	$\Delta P_{I,b_{O_2}}$ (Torr)	$P_{b_{CO_2}}$ (Torr)	pHb	$[HCO_3^-]b$ (meq · L^{-1})
1	Dogfish,	Gill	16	149	56	93	2.0	7.78	
	Scyliorhinus stellaris		*25*				*3.4*	*7.63*	
11	Garfish,	Lung, gill	25	155			13	7.44	10.2
	Lepisosteus osseus								
4	Electric eel,	Buccal mucosa, skin	28	150	12	138	29	7.58	30
	Electrophorus electricus		*25*				*24.2*	*7.63*	
10	*Amphipnous cuchia*	Air sacs, gill, skin	25	155			12	7.50	
8	Australian lungfish,	Gill (lung)	18	131	40	91	3.6	7.64	5.3
	Neoceratodus forsteri		*25*				*5.7*	7.56	
9	African lungfish,	Lung (gill)	25	130	27	103	26		
	Protopterus aethiopicus								
3	Bullfrog, *Rana catesbeiana*								
	tadpole	Gill, skin	20	155			3.2	7.83	5.5
			25				*4.3*	*7.75*	
	adult	Skin, lung	20	155			13	7.90	30
			25				*17.5*	*7.82*	

7	Bullfrog, *Rana catesbeiana*								
	tadpole	Gill, skin	23	145			4.5	7.80	8.0
			25				*5.1*	*7.77*	
	adult	Skin, lung	23	145			18	7.68	28
			25				*20.3*	*7.65*	
2	Hellbender salamander, *Cryptobranchus allegan.*	Skin (lung)	25	154	27	127	6.2	7.78	
5	Dusky salamander,	Skin	13	150	40	140	6.2	7.51	9.15
	Desmognathus fuscus		*25*				*9.5*	*7.32*	
12	Toad adult, *Bufo marinus*	Skin, lung	25	155	80	75	11.1	7.82	21.4
6	Red-eared turtle,	Lung	20	155			22.7	7.67	34.8
	Pseudemys scripta el.		*25*				*27.5*	*7.61*	

Columns 6–10 concern blood (see table 4.2).
Data for the dogfish and for the red-eared turtle are drawn from tables 4.1 and 4.2 respectively.

1 Baumgarten-Schumann and Piiper (1968)
2 Boutilier and Toews (1981)
3 Erasmus *et al.* (1970)
4 Garey and Rahn (1970)
5 Gatz *et al.* (1974)
6 Jackson *et al.* (1974)
7 Just *et al.* (1973)
8 Lenfant *et al.* (1966)
9 Lenfant and Johansen (1968)
10 Lomholt and Johansen (1976)
11 Rahn *et al.* (1971)
12 Toews and Heisler (1982)

TABLE 4.5

Acid–base balance of the prebranchial hemolymph in the shore crab, *Carcinus maenas* (Truchot, 1979) and in the freshwater crayfish, *Austropotamobius pallipes* (Taylor and Wheatly, 1980), immersed in normoxic water or emersed.

	Crab (16 °C)		Crayfish (15 °C)	
	immersed	emersed (2.25 h)	immersed	emersed (3 h)
pH	7.77	7.81	7.86	7.43
P_{CO_2} (Torr)	1.6	5.2	3.5	8.9
[Carbonates] ($meq \cdot L^{-1}$)	4.47	15.5		

water, breathes air from time to time, and accumulates CO_2. However, the accretion of CO_2 in the body is accompanied by an increase of P_{CO_2}, so that some CO_2 leaks through the skin into the water, as long as the water P_{CO_2} is low, or into the air if the animal is completely emersed but wet.

Other examples of ABB dependency on the nature of the environment are given by the study of the intertidal shore crab (Truchot, 1979) and the freshwater crayfish (Taylor and Wheatly, 1980). Most of the time these crustaceans live in water, but the crab can be emersed at low tide and the crayfish may emerse from time to time to search for food or to colonize new bodies of water. During emersion, the ventilation of the branchial cavities decreases. Table 4.5 shows the ABB of the prebranchial hemolymph in both species; emersion (2–3 h) leads to a hypercapnia, which disappears very quickly when the animals are reimmersed. Note that some crabs (Truchot, 1987a; Burnett and McMahon, 1987), but not the crayfish, compensate the hypercapnic acidosis (p. 99).

These tables and figures summarize most of the available data, and most of the conclusions are probably valid. More observations will be welcome, above all in experiments where the ambient conditions, particularly of the acid–base balance prevailing in the water, are specified and, better controlled.

CHAPTER 5

Responses to ambient oxygen and carbon dioxide variations

Summary

A normoxic milieu is by definition one whose O_2 tension is that of air at sea level, about 150–160 Torr. For a given P_{O_2}, air always contains much more oxygen than water. In normoxic *water breathers*, hyperoxia induces a very marked fall of the ventilation and hypoxia a marked increase. The hyperoxia-induced hypoventilation and the hypoxia-induced hyperventilation entail, respectively, hypercapnia and acidosis, hypocapnia and alkalosis. In normoxic *air breathers*, hyperoxia in some species leads to a moderate fall of ventilation with a modest hypercapnia. Hypoxia may result in a very marked hyperventilation with important hypocapnia and alkalosis. Changes of ambient oxygenation affect water- and air breathers' ventilation differently (fig. 5.4). The higher the arterial blood O_2 concentration, the lower the cardiac output, with no difference between air- and water breathers (fig. 5.5).

Ambient hypercapnia in water breathers induces a small increase of ventilation or none. Ambient hypercapnia, in normoxic air breathers, increases ventilation, a response exacerbated by hypoxia, except at very low PI_{O_2}, such as that at very high altitude.

In air-breathing vertebrates, the ventilatory responses to changes of the ambient oxygenation are mainly, if not exclusively, reflexly driven by the stimulation of arterial chemoreceptors (Heymans-type chemoreceptors). The ventilatory adaptation to high altitude depends on these chemoreceptors. Their destruction results in a ventilatory insufficiency. Apparently, but not as yet surely, analogous chemoreceptors exist in some fish, crustaceans and annelids.

The ventilatory responses to changes of ambient carbon dioxide are more complex. The several possible mechanisms through various anatomical structures include central (brainstem) chemoreceptors; Heymans-type chemoreceptors; special intrapulmonary CO_2 receptors (IPC) of reptiles and birds and pulmonary stretch receptors modulated by the CO_2 tension in reptiles. All four types of receptors have been seen only in reptiles. The physiological role of IPC and CO_2-modulated mechanoreceptors is not clear.

In conclusion, water breathers are mainly sensitive to changes of the oxygenation of their medium. Air breathers respond to changes of both O_2 and CO_2 ambient pressures, except extremely hypoxic animals which hyperventilate irrespective of arterial hypocapnia. All ventilatory reactions to changes of environmental O_2 and CO_2 are viewed as tending to ensure proper oxygenation of the body, and to a certain extent a proper acid–base balance.

Animal life requires the uptake of oxygen and the clearance of carbon dioxide. How do organisms react to variations of the oxygen and carbon dioxide characteristics of the milieu? By characteristics I mean the relationship $C_x = \beta_x \cdot P_x$ (Chapter 3). For oxygen the relation is simple, because the capacitance (solubility) for oxygen is well defined and constant for a given milieu. For carbon dioxide, the relation is complex, as explained in Chapter 3.

There was a time when carbon dioxide was thought of as the most important stimulus of external respiration. This conviction came from the study of mammals, of man in particular, at sea level, in normoxic conditions, and it is wrong. Breathing the modern atmosphere at sea level is the result of a long evolution. Comparative physiological experiments show that oxygenation is the primary goal of respiration and that carbon dioxide clearance is subordinate, and that in case of conflict between a proper oxygenation and a proper CO_2 clearance, oxygenation takes precedence.

Ambient oxygen variations

A water is normoxic when it is in equilibrium with air at sea level, that is at a partial pressure of 150–160 Torr, the actual value depending on water temperature. But for the same normoxic P_{O_2} value, the oxygen concentration C_{O_2} may decrease considerably with a rise in temperature or salinity or both (Chapter 3). Normoxia, as defined above, could be called a *baronormoxia*, keeping in mind that the prevailing O_2 concentration in a baronormoxic condition varies with the characteristics of the milieu.

External respiration

Water breathers

Table 5.1 shows the qualitative variations of ventilation induced by aquatic hypoxia or hyperoxia of some selected water breathers belonging

TABLE 5.1

Action of hypoxia, hyperoxia and normoxic hypercapnia on ventilation ($\dot{V}$) in seawater and freshwater animals. For a more complete list of references, see Wilkes *et al.* (1981) and Shelton *et al.* (1986).

Ref.	Species	Method	Hypoxia	Hyperoxia	Hypercapnia in normoxia
	ANNELIDS				
1, 1b	*Arenicola marina*	collection of expired water	increases $\dot{V}$	decreases $\dot{V}$	no response
	CRUSTACEANS				
2	*Carcinus maenas*	collection of expired water	increases $\dot{V}$	decreases $\dot{V}$	no response or small increase
3	*Astacus leptodactylus*	collection of expired water	increases $\dot{V}$	decreases $\dot{V}$	variable increase of $\dot{V}$
	MOLLUSCS				
4	*Anondonta cygnea*	visual observ. polarogram	increases $\dot{V}$	decreases $\dot{V}$	
	FISHES				
5	*Anguilla vulgaris*	collection of expired water	increases $\dot{V}$		small effect
	Salmo shasta				increases $\dot{V}$
6	*Catostomus commersoni*	Saunders' method	increases $\dot{V}$		small increases of $\dot{V}$
	Ictalurus nebulosus	Saunders' method	increases $\dot{V}$		small increases of $\dot{V}$
	Cyprinus carpio	Saunders' method	increases $\dot{V}$		small increases of $\dot{V}$
7	*Cyprinus carpio*	opercular mechanogram	increases $\dot{V}$	decreases $\dot{V}$	small effect
8	Several Cyprinidae	modified Saunders' method	increases $\dot{V}$	decreases $\dot{V}$	small effect
	Salmo trutta	modified Saunders' method	increases $\dot{V}$	decreases $\dot{V}$	increases $\dot{V}$
9	Several marine fishes	opercular water sampling		decreases $\dot{V}$	
	AMPHIBIANS				
10	Tadpole *Rana catesbeiana*	expired water collection	increases $\dot{V}$	decreases $\dot{V}$	

1 Toulmond and Tchernigovtzeff (1984)
1b Conti and Toulmond (1986)
2 Jouve-Duhamel and Truchot (1983)
3 Massabuau *et al.* (1984)
4 Koch and Hers (1943)
5 Van Dam (1938)
6 Saunders (1962)
7 Peyraud and Serfaty (1964)
8 Dejours (1973)
9 Dejours *et al.* (1977)
10 West and Burggren (1982)

to a few phyla. It is well known that hypoxia increases ventilation in all animals studied. The ventilatory depression induced by hyperoxia has been less well documented (Dejours, 1973), but all observations show it to be a general phenomenon (see Toulmond and Tchernigovtzeff, 1984). Note that aquatic hyperoxia is a natural phenomenon (see p. 26).

In many animals, belonging to different phyla, marked hypoxia or near complete anoxia leads to a reversible arrest of respiration. This is quite clearly observed in the lugworm, which is exposed to anoxia at low tide (Toulmond and Tchernigovtzeff, 1984), and in the prawn, which may become hypoxic at night in a rockpool (Morris and Taylor, 1985).

In water breathers, such as many molluscs, crustaceans, fishes, and in air breathers such as reptiles, birds and mammals, all animals with a breathing apparatus providing an ingoing and an outgoing flow of ambient medium and having a negligible cutaneous respiration, the O_2 consumption in steady state may be expressed by the following relationship:

$$\dot{M}_{O_2} = \dot{V} \cdot (C_{I_{O_2}} - C_{E_{O_2}}) \tag{5.1}$$

A simple derivation (see Dejours *et al.*, 1970) leads to the following:

$$\dot{M}_{O_2} = \dot{V} \cdot E \cdot C_{I_{O_2}} \tag{5.2a}$$

which for water- and air breathers respectively is

$$\dot{M}_{O_2} = \dot{V}w \cdot Ew \cdot C_{I_{O_2}} \tag{5.2b}$$
$$\dot{M}_{O_2} = \dot{V}g \cdot Eg \cdot C_{I_{O_2}} \tag{5.2c}$$

where E denotes the extraction coefficient of O_2 from the milieu. In order to compare the ventilation of animals of different groups and sizes and at various levels of activity, due, *e.g.* to ambient temperature changes, the ratio $\dot{V}w \cdot \dot{M}_{O_2}^{-1}$ and $\dot{V}g \cdot \dot{M}_{O_2}^{-1}$ (p. 32), the specific ventilation, is used.

Figure 5.1 gives selected examples of $\dot{V}w \cdot \dot{M}_{O_2}^{-1}$ changes in three species of water breathers as a function of $C_{I_{O_2}}$; the top of fig. 5.2 shows another example in the crayfish as a function of $P_{I_{O_2}}$.

A derivation of eq. 5.1 leads to

$$\dot{M}_{O_2} = \dot{V} \cdot \beta_{O_2} \cdot (P_{I_{O_2}} - P_{E_{O_2}}) \tag{5.3}$$

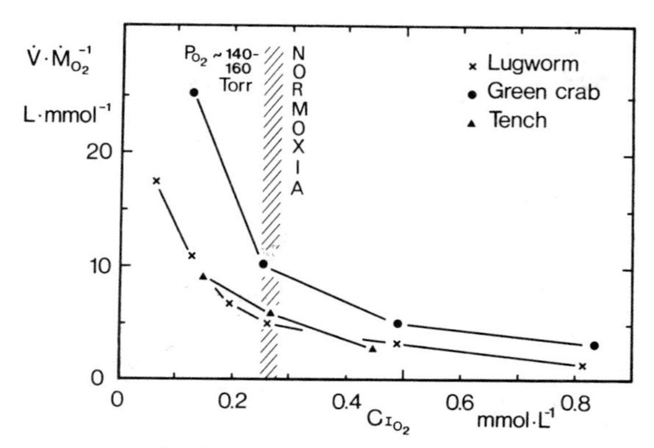

Fig. 5.1. Specific ventilation, $\dot{V} \cdot \dot{M}_{O_2}^{-1}$, as a function of the O_2 concentration of the inhaled water, $C_{I_{O_2}}$, in the lugworm *Arenicola marina* (Toulmond and Tchernigovtzeff, 1984), the green crab *Carcinus maenas* (Jouve-Duhamel and Truchot, 1983) and the tench *Tinca tinca* (Dejours, 1973). The temperature is 15 °C for the lugworm and the crab, and 20 °C for the tench. The dashed band is the normoxic zone, $P_{O_2} \simeq$ 140–160 Torr.

where β_{O_2} denotes the capacitance coefficient of O_2 in a given medium.

Similarly, for carbon dioxide,

$$\dot{M}_{CO_2} = \dot{V} \cdot \beta_{CO_2} \cdot (P_{E_{CO_2}} - P_{I_{CO_2}}) \tag{5.4}$$

The examination of eq. (5.3) shows that at constant $\dot{M}_{O_2}$, any change in $\dot{V}$ is coupled with a reciprocal change of $(P_{I_{O_2}} - P_{E_{O_2}})$. The same can be said for the difference $(P_{I_{O_2}} - P_{I_{CO_2}})$ of eq. (5.4) with the reserve that β_{CO_2} in water may not be constant (see p. 20). Anyway, any decrease of $\dot{V}$ leads to increases of $(P_{E_{CO_2}} - P_{E_{O_2}})$ and $(P_{E_{CO_2}} - P_{I_{CO_2}})$. Although the relation between $(P_{E_{CO_2}} - P_{I_{CO_2}})$ and body fluids, *e.g.* arterial P_{CO_2}, is complex; particularly in aquatic animals, any increase of $(P_{E_{CO_2}} - P_{I_{CO_2}})$ leads to a hypercapnia in body fluids.

When ventilation is depressed by hyperoxia, the differences $(P_{I_{O_2}} - P_{E_{O_2}})$ and $(P_{E_{CO_2}} - P_{I_{CO_2}})$ do actually increase, and a hypercapnia and an acidosis are observed. In animals whose ventilation is enhanced by hypoxia, the reverse effect, hypocapnia and hypocapnic alkalosis, is observed. Figure 5.2 shows an example of the effect of the change in ambient P_{O_2} on $\dot{V} \cdot \dot{M}_{O_2}^{-1}$ and on P_{CO_2}, pH and $[HCO_3^-]$ of the prebranchial blood of the crayfish. I do not know of any exception to these modifications in specific ventilation and acid–base balance brought about by variations of the ambient oxygenation. It has sometimes been said that

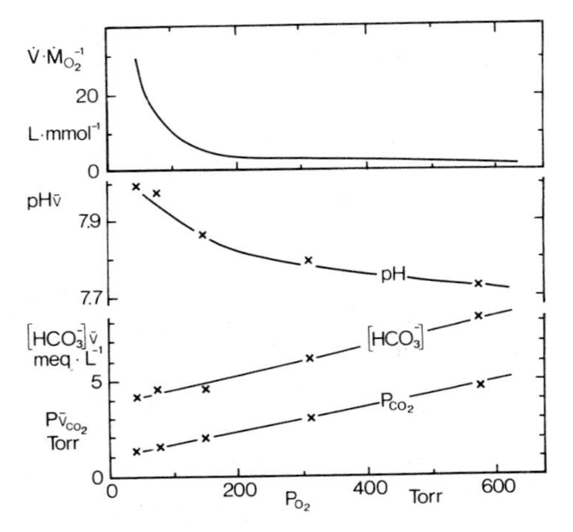

Fig. 5.2. Specific ventilation, $\dot{V} \cdot \dot{M}_{O_2}^{-1}$ and pH, $[HCO_3^-]$ and P_{CO_2} of prebranchial (mixed venous) hemolymph in 17 crayfish *Astacus leptodactylus* 24 h after exposure to various water oxygen tensions. Ambient temperature was 13 °C and the water pH and P_{CO_2} were regulated at 8.40 and 0.8 Torr (from Sinha and Dejours, 1980).

hypoxic hyperventilation does not lead to hypocapnia, because the arterial P_{CO_2} value is already very low, but when the ambient conditions are well controlled, and blood carefully sampled and analyzed, a hypocapnic alkalosis is always observed (see Truchot, 1987).

Animals differ, however, in their capacity to compensate hyperoxia-induced hypercapnic acidosis and hypoxia-induced hypocapnic alkalosis. Figure 5.2 shows that in the crayfish *Astacus leptodactylus* 24 h after exposure to a change of oxygenation, the acidosis of hyperoxia and the alkalosis of hypoxia persist, and the analysis of the data show that respiratory ABB changes are only slightly compensated metabolically. In this species, even after several weeks of hyperoxia-induced hypercapnia in the particular conditions of the experiment, the acidosis persists (Dejours and Beekenkamp, 1977). In the freshwater white sucker, Wilkes *et al.* (1981) observed a partial compensation of the hypercapnic acidosis after 3 days of exposure to hyperoxia. In the rainbow trout, the compensation was complete within 48 h (Wood and Jackson, 1980). In the shore crab (Truchot, 1975), the deviations of the blood ABB due to environmental changes of the oxygenation are almost completely compensated (fig. 8.3). In the crab *Cancer productus*, DeFur *et al.* (1980) suggested that $CaCO_3$, possibly from the exoskeleton, could be mobilized during the acid–base compensation. Truchot (1987) gives a detailed review of the literature on this problem in various molluscs, crustaceans and fishes.

Air breathers

The variations of ventilation as a function of ambient P_{O_2} have been extensively studied in vertebrates (see summary by Bouverot, 1985).

By convention, the ventilation at sea level is taken as the reference value. Prolonged hyperoxia, let us say 300–400 Torr, leads to no change in ventilation in humans, but causes a sustained hypoventilation in dogs, rats and chickens (Bouverot, 1985). All air breathers studied differ from water breathers, vertebrates or not, by the fact that the hyperoxia-induced depression of ventilation is much less pronounced in normoxic air breathers than in normoxic water breathers.

In air breathers, hypoxia entails an increase of ventilation which may be extremely high in marked hypoxia (fig. 5.3). Obviously such an increase of ventilation leads to a hypocapnia and an alkalosis. If the altitude is not too high, the hypocapnic alkalosis may be compensated by a decrease in $[HCO_3^-]$, but at very high altitude, as for humans on the upper slopes of Mount Everest, a marked alkalosis was predicted (Dejours, 1979, 1981, 1982) and was actually observed by West and his coworkers during the 1981 expedition to Everest (West, 1983; West and Lahiri, 1984). The lowest value of alveolar P_{CO_2} recorded there was 7.5 Torr (West, 1983).

Air breathers compared to water breathers

Figure 5.4 shows the data of figs. 5.1 and 5.3 for water and air breathers plotted to the same scale. The differences between air- and water breathers

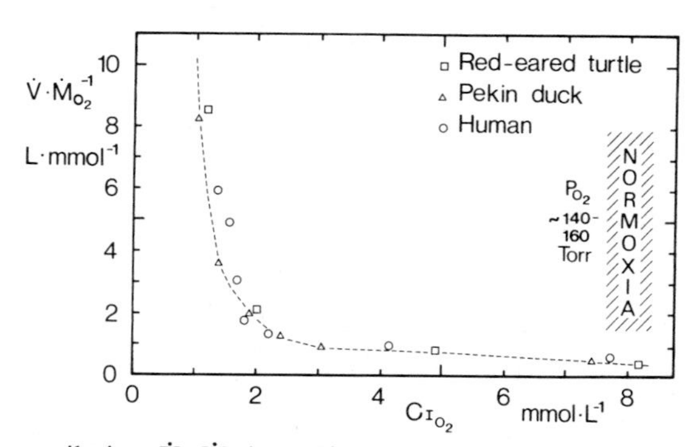

Fig. 5.3. Specific ventilation, $\dot{V} \cdot \dot{M}_{O_2}^{-1}$, as a function of the O_2 concentration of the inhaled air, $C_{I_{O_2}}$, in a freshwater turtle at 30 °C (Jackson, 1973), the Pekin duck acclimated at 5640 m (Black and Tenney, 1980) and the human acclimated at 4540 m (Velasquez, 1959). The vertical dashed band indicates normoxia, $P_{O_2} \simeq$ 140–160 Torr.

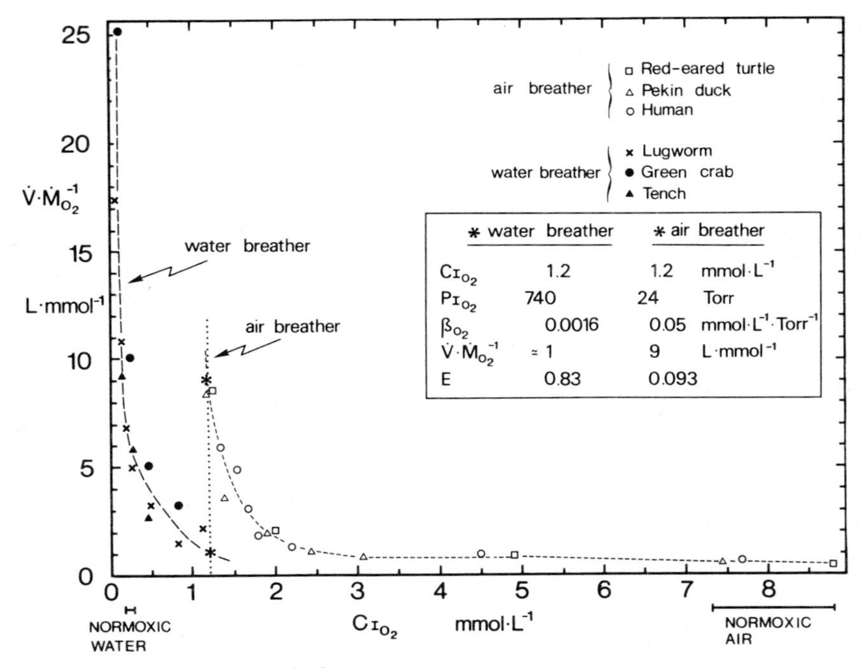

Fig. 5.4. Specific ventilation, $\dot{V}/\dot{M}_{O_2}$, as a function of the inspired O_2 concentration, CI_{O_2}, in water- and air breathers. This figure recapitulates the data of figs. 5.1 and 5.3 but uses the same coordinates for both groups of animals. The horizontal bars under the abscissa scale indicate the baronormoxic range for each group. Their widths take into account the pressure range of 140–160 Torr, and the variation of the solubility coefficient and the water vapor change with temperature. The intercepts of the vertical dashed line at abscissa CI_{O_2} = 1.2 mmol · L^{-1} with the water breathers' and air breathers' lines are marked by asterisks. The insert shows some characteristics of these intercepts. E designates the O_2 extraction coefficient.

are striking. One may see that for each group in normoxia, indicated by horizontal bars below the abscissa, the specific ventilation is much higher in water breathers than in air breathers. At the same value of CI_{O_2} (1.2 mmol · L^{-1}), the theoretical points (asterisks) on the water breathers' and air breathers' lines differ markedly: the specific ventilation is much higher in the air breather than in the water breather. It is presumably so high in the air breathers because the oxygen in air is there at a very low partial pressure, a pressure head which has to diminish at each step of the respiratory system from the air down to the mitochondria. It may mean that the O_2 partial pressure is more important than the actual O_2 concentration in driving the air-breathers' ventilation, a point that should be explored. On the other hand, the high specific ventilation of the very hypoxic air breathers is possible because air is a much less dense and viscous milieu than water.

In the general eq. (5.2a)

$$\dot{M}_{O_2} = \dot{V} \cdot E \cdot CI_{O_2}$$

CI_{O_2}, the O_2 concentration in the milieu, is an environmental parameter to which the organism is exposed; on the other hand, the product $\dot{V} \cdot E$, that is the product of the ventilatory flow rate times the extraction coefficient, is the expression of the external respiration. Since for the same CI_{O_2}, 1.2 mmol $\cdot$ L^{-1}, the specific ventilation is 9 times higher in the air breather than in the water breather, it follows that the air breather's extraction coefficient is 9 times lower than in the water breather. As a general rule, but with considerable scattering, one may say that the higher the environmental O_2 concentration, the lower the extraction coefficient (Dejours *et al.*, 1970). This is quite clear in the insert of fig. 5.4 and in table 4.1 (line q).

Cardiac output

In animals having a well-defined arterialized blood and no important cutaneous O_2 and CO_2 exchanges, the Fick equation relates the oxygen consumption, $\dot{M}_{O_2}$, the blood flow rate, $\dot{V}b$, and the arterial–venous oxygen concentration difference, $Ca_{O_2} - C\bar{v}_{O_2}$,

$$\dot{M}_{O_2} = \dot{V}b \cdot (Ca_{O_2} - C\bar{v}_{O_2}) \quad (5.5)$$

which can be transformed to

$$\dot{M}_{O_2} = \dot{V}b \cdot Eb_{O_2} \cdot Ca_{O_2} \quad (5.6)$$

in which Eb_{O_2} denotes the extraction coefficient of oxygen from the arterial blood.

Equation 5.6 may be written

$$\frac{\dot{V}b}{\dot{M}_{O_2}} \cdot \frac{1}{Eb_{O_2} \cdot Ca} = 1 \quad (5.7)$$

$\dot{V}b/\dot{M}_{O_2}$ being the specific blood flow.

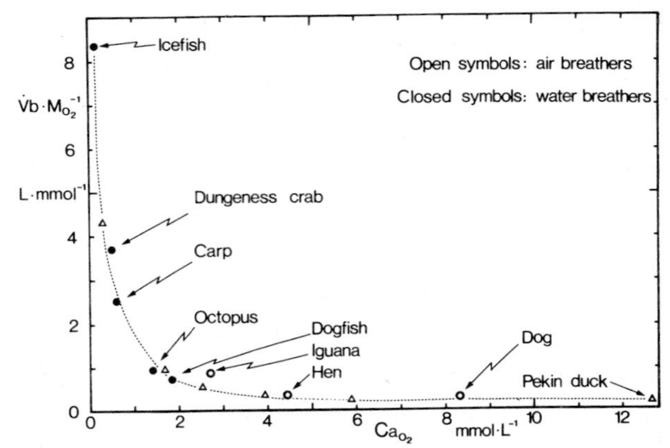

Fig. 5.5. Specific blood flow, $\dot{V}b \cdot \dot{M}_{O_2}^{-1}$, as a function of the arterial blood O_2 concentration, Ca_{O_2}, in some water- and air breathers. The Pekin duck points are taken from Black and Tenney (1980); the other points are reproduced from Dejours (1981, p. 120). It may be seen that water and air breathers do not differ. The extraction coefficients are the highest in animals whose arterial C_{O_2} is low: 0.50 in the duck with $Ca_{O_2} = 12.6\ mmol \cdot L^{-1}$, and 0.80 in the hemoglobinless icefish in which all arterial oxygen is in solution, $Ca_{O_2} = 0.15\ mmol \cdot L^{-1}$.

Let us examine how $\dot{V}b \cdot \dot{M}_{O_2}^{-1}$ changes with the very large variation of Ca_{O_2} (fig. 5.5). Some animals like mammals and birds, mainly those adapted to very high altitude, may have a very high concentration of O_2 in the arterial blood, related to their hypoxia-induced hyperhemoglobinemia (the highest Ca_{O_2} in fig. 5.5 is 12.6 mmol $\cdot$ L^{-1} in the high-altitude adapted duck measured at sea level; Black and Tenney, 1980). Other animals have a low concentration of O_2-carrying pigment or no pigment at all, as the hemoglobinless icefish, in which all blood oxygen is in solution (the lowest Ca_{O_2} in fig. 5.5 is the icefish's 0.15 mmol $\cdot$ L^{-1}). Since the ratio of Ca_{O_2} between these two examples is about 100, it follows that to satisfy eq. (5.7), $\dot{V}b \cdot \dot{M}_{O_2}^{-1}$ decreases considerably as Ca_{O_2} increases, a relation shown in fig. 5.5. There is also a diminution of the extraction coefficient Eb_{O_2} with the rise of Ca_{O_2}. These observations are well established (see Dejours *et al.*, 1970; Mangum, 1977). As long as the anatomical disposition of the circulatory system is such that eq. (5.5) can be written, eqs. (5.6) and (5.7) must be fulfilled.

It is certainly not with the help of the $\dot{V}b \cdot \dot{M}_{O_2}^{-1} = f(Ca_{O_2})$ relationship that one can distinguish air breathers from water breathers: as fig. 5.5 shows, they are exactly the same.

Ambient carbon dioxide variations

Water breathers

Table 5.1 shows that ambient hypercapnia does not cause much increase of the ventilation of water breathers. When one wants to study the possible action of ambient hypercapnia, the CO_2-enrichment of the water to be inhaled should be limited to a few torrs of CO_2, such as may occur in natural conditions. A marked hypercapnia will not mean much as to the regulation of breathing, since it may very well depress ventilation by inducing a narcotizing acidosis, an effect for which it is sometimes used purposely.

Hypercapnia does, however, induce some hyperventilation in several normoxic water breathers. Massabuau *et al.* (1984b, 1985) have suggested that water breathers whose ventilation responds to hypercapnia are only those with a relatively high energy metabolism, but whether this is a general rule is unknown. We have seen that hyperoxia always results in hypoventilation and hypercapnia. I have argued that this hyperoxia-induced hypercapnia implies that the hyperoxic animal is insensitive to hypercapnia, and most probably to ambient hypercapnia (Dejours, 1973). Indeed, in most experiments, hyperoxia depresses considerably or entirely the ventilatory response to hypercapnia. It is the case in various Ciprinidae, in trout (Dejours, 1973), in the crayfish *Astacus leptodactylus* (Massabuau *et al.*, 1984), and in the green crab (Jouve-Duhamel and Truchot, 1983). Hypercapnia's lack of effect on ventilation in hyperoxic water breathers may seem rationally obvious, but it had to be tested experimentally. Thomas *et al.* (1983), however, reported a marked increase in ventilation in the rainbow trout made hypercapnic, *Salmo gairdneri*, even under hyperoxia, an observation I have never been able to duplicate. Until further experiments are reported, one may consider that water breathers are little sensitive to ambient hypercapnia.

Air breathers

Air breathers in mild hypoxia, normoxia or hyperoxia, inhaling CO_2-enriched mixtures, increase their ventilation. The pulmonary and arterial hypercapnia are then moderated, as initially reported by Haldane and Priestley (1905). The literature is abundant on this subject.

There are three circumstances in which the ventilatory responsiveness to CO_2 is small or absent.

(1) In fossorial mammals and birds, hypercapnia-induced hyperventilation is less than the response observed in open-air mammals (see Boggs *et al.*, 1984; Tenney, 1986).

(2) In diving mammals and birds, the ventilatory response to carbon dioxide inhalation is less marked than in non-diving animals, for instance the harbor seal (Robin *et al.*, 1963; Tenney, 1986).

(3) Extreme hypoxia. Nielsen and Smith (1951) studied two human subjects at various levels of ambient oxygenation. A very low oxygen ambient pressure (alveolar P_{O_2} at either 47 or 37 Torr) entails a very important hyperventilation with a marked hypocapnia (fig. 5.6), a phenomenon which has been widely confirmed (*e.g.* Velasquez, 1959). The originality of the paper by Nielsen and Smith is that they studied the ventilatory response to the inhalation of hypercapnic mixtures in very hypoxic subjects.

Figure 5.6 shows the results obtained in one subject. In the normoxic or slightly hyperoxic subject (curve A) hypercapnia induced hyperventilation. In marked hypoxia (curve B, PA_{O_2} = 47 Torr; curve C, PA_{O_2} = 37

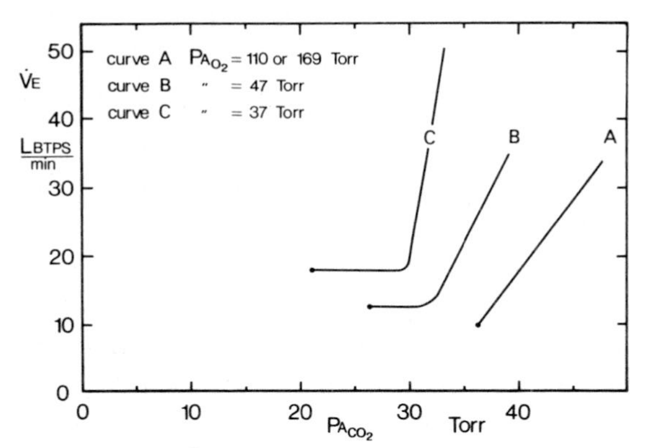

Fig. 5.6. Ventilatory flow rate, $\dot{V}_E$, as a function of alveolar P_{CO_2}, PA_{CO_2}, in one human subject breathing air or a slightly hyperoxic gas (curve A), or a hypoxic mixture with PA_{O_2} equal to 47 Torr (curve B) or 37 Torr (curve C). The point at the lower left of each curve: the subject breathed a gas mixture with a negligible amount of CO_2. To obtain the solid part of each curve, the subject was given a CO_2-enriched mixture to breathe. In the hypoxic conditions B and C, a very marked increase of ventilation occurred only above a certain PA_{CO_2} threshold (redrawn from Nielsen and Smith, 1951).

Torr), there was a P_{CO_2} threshold value, 30–32 Torr, above which a hyperventilatory response to CO_2 was seen. In curve C, the $P_{A_{CO_2}}$ value of the subject breathing 7.9% O_2 in N_2 at sea level was 21 Torr. When the subject was given CO_2-containing mixtures to breathe, the pulmonary ventilation remained unchanged until $P_{A_{CO_2}}$ was about 30 Torr; above that value the ventilation increased steeply. In short, hypoxic subjects are insensitive to variations of blood P_{CO_2} (and pH) below a certain level of P_{CO_2}. The slope of the ventilatory CO_2 response curve, an index of ventilatory CO_2 sensitivity, was the steeper the greater the hypoxia. The two stimuli, hypercapnia and hypoxia, interact positively.

The data on the CO_2 sensitivity of very hypoxic subjects are scarce, presumably because such a degree of hypoxia may not be harmless. The effect has, however, been confirmed in goats (Lahiri *et al.*, 1971). One may look at the insensitivity of very hypoxic subjects to variations of P_{O_2} (below a certain CO_2 threshold) in a teleonomic way. Obviously if the subjects were still sensitive to hypocapnia below 30 Torr, they would breathe less, would become still more hypoxic, and their life would be endangered.

Figure 5.6, concerning one human subject in normoxia (A), shows that his ventilatory response as a function of alveolar P_{CO_2} is a straight line, but more often the response to hypercapnia is curvilinear (fig. 5.7). The slope, the index of CO_2 responsiveness, is rather low near the normocapnic value and increases progressively to become approximately constant in marked

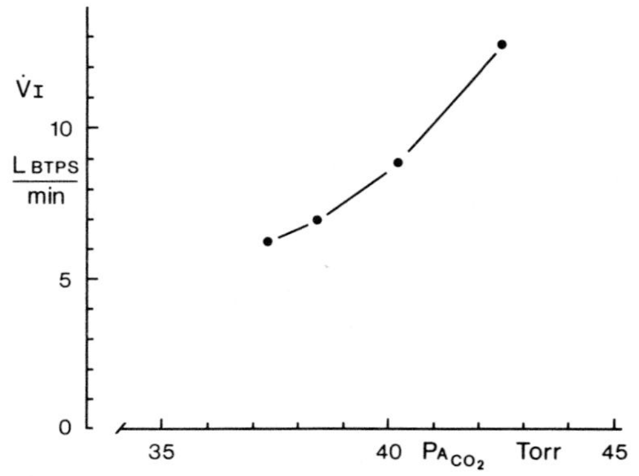

Fig. 5.7. Inspired ventilatory flow rate, $\dot{V}_I$, as a function of alveolar P_{CO_2}, $P_{A_{CO_2}}$, between the 20th and 25th minutes of inhaling ambient air and 0.010, 0.025 and 0.040 CO_2 diluted in air. Means of 4 subjects (from Dejours *et al.*, 1965).

hypercapnia (Anthonisen *et al.*, 1965; Dejours *et al.*, 1965; Forster *et al.*, 1982), as well in awake as in sleeping subjects (Reed and Kellogg, 1960).

The regulation of ventilation by the CO_2 tension is generally considered as being proportional (p. 14). Anyway, if an integral control is associated with a proportional control, an excellent regulation with a very small change in blood P_{CO_2} may be observed (Coon *et al.*, 1984). Possibly the CO_2 stores in the brain stem may play the role of an integral controller.

Ventilatory response mechanisms to ambient changes of oxygen and carbon dioxide

Figure 5.8 shows diagrammatically the ventilatory neuromechanical system and the various afferent inputs which can modify its activity. The neuromechanical system is a functional unit with the respiratory centers (RC) connected by motoneurons to the respiratory apparatus whose disposition varies with the zoological groups: lung and chest in most reptiles, birds and mammals; lung and the complex cephalic muscles in amphibians; gills and the bucco-opercular pump in fish, and gill and cilia, scaphognathites and body muscles in invertebrates.

In fish, reptiles, birds and mammals, nerves originating in the respiratory apparatus send information to the respiratory centers (Fedde and Kuhlmann, 1978; Ballintijn, 1982). In their study of awake dogs, Kelsen *et al.* (1982) concluded that the integrity of the neuromechanical system, a closed loop, seems to be important in reducing the breath-to-breath scatter of the respiratory variables: ventilation and alveolar P_{CO_2}. The neurochemical functional unit in invertebrates may not be a closed loop, since an afferent pathway has not been demonstrated.

The upper part of fig. 5.8 classifies stimuli as neural and humoral. The legend gives more details. In this chapter we will deal only with certain principles concerning the stimulation of breathing by the oxygenation and 'carboxygenation' of the milieu and the body fluids and with some problems they raise. In Chapter 6, other stimuli which account for respiratory control under a heat load and in exercise will be discussed.

Ambient oxygen

A decrease of ambient P_{O_2} induces a fall of P_{O_2} in body fluids. Below a

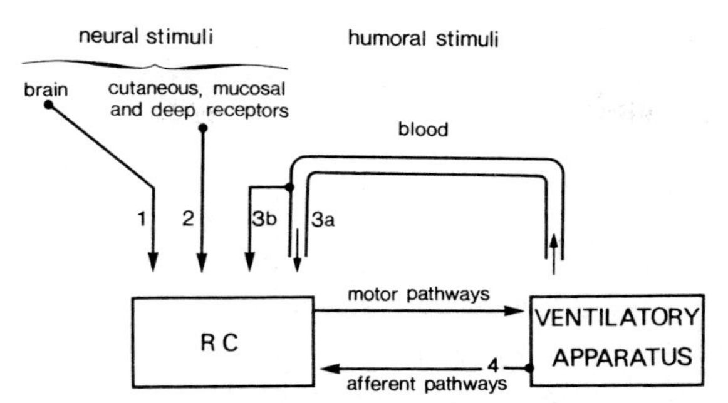

Fig. 5.8. The ventilatory neuromechanical system, a single functional unit composed of the respiratory centers (RC), the motor pathways to the respiratory muscles, the respiratory apparatus, and the afferent pathways connecting this apparatus to the RC. The RC receive information on bodily status and requirements through stimuli that can be classified as neural stimuli or humoral (blood-borne) stimuli (from Dejours, 1966).

Neural stimuli (pathways 1, 2 and 4)	Intercentral (pathway 1)	cortical or subcortical origin
	Reflex (pathways 2 and 4)	cutaneous receptors mucosal receptors proprioceptors (somatic mechanoreceptors) pulmonary or branchial chemoreceptors or mechanoreceptors
Blood-borne stimuli (pathway 3)	Chemical	hypoxia (pathway 3b) hypercapnia, acidosis (pathways 3a and b) catecholamines other chemical factors
	Physical	blood temperature blood pressure (pathway 3b)

certain value, a threshold, the arterial P_{O_2} stimulates chemoreceptors. These are sensory cells which are usually grouped near the aortic arch in most mammals (see Heymans and Neil, 1958). Birds (see Bouverot, 1978) and certain reptiles (Ishii *et al.*, 1985) also have grouped chemoreceptors perfused by arterialized blood. They are stimulated to fire by a drop in the prevailing blood P_{O_2}, and their afferent innervating nerves send impulses towards the brain stem (pathway 3b of fig. 5.8). The RC are stimulated and the ventilation increases. This is a hypoxia-induced chemoreflex drive of ventilation.

Thus the complete demonstration of a chemoreflex mechanism requires three lines of evidence, namely (1) morphological: chemoreceptor cells, chemoreceptor organ, innervation; (2) electrophysiological: discharge along the afferent fibers innervating the chemoreceptors, and stimulus–response curves; and (3) physiological: a ventilatory effect which increases with the intensity of the chemoreflex stimulus and which disappears if the nervous connection between chemoreceptor body and the brain stem is blocked by cold or by section.

The demonstration of a chemoreflex drive is not easy in intact animals. It cannot be deduced from the changes of ventilation related to *prolonged*

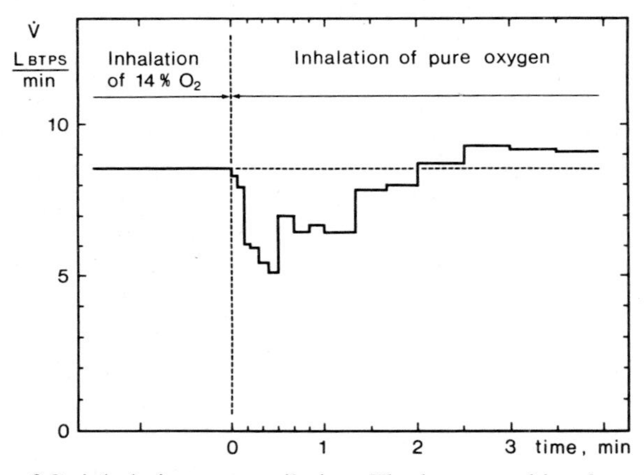

Fig. 5.9. Effects of O_2 inhalation on ventilation. The human subject breathed a hypoxic mixture containing 14% O_2 in nitrogen for a few minutes. Starting at time 0, the subject inhaled pure O_2. Ventilation was measured for each cycle during the first 30 sec and then for groups of several cycles. A few seconds after O_2 breathing started, ventilation decreased markedly. The O_2 inhalation abolished the hypoxic stimuli present while a hypoxic mixture was inhaled. The hypoxic stimulus acts indirectly on ventilation, by a reflex mechanism originating in the aortic and carotid Heymans-type receptors. When inhalation of pure O_2 was maintained, ventilation increased and even exceeded the ventilation of the hypoxic period. The mechanism controlling ventilation during prolonged O_2 breathing is complex. Not only the O_2 stimuli, but other factors of regulation of ventilation are modified. In particular, the respiratory centers become hypercapnic (see Lambertsen *et al.*, 1953). The secondary modifications due to hyperoxygenation mask the ventilatory effects due to abolition of the hypoxic stimulus, effects visible only during the initial phase of O_2 inhalation, before the secondary effects have had time to appear (from Dejours, 1957, 1981).

changes of the body oxygenation, because many factors other than the chemoreflex drive then come into play, as explained in the legend of fig. 5.9. But the study of the effect of an abrupt variation of inspired P_{O_2} on the immediately following ventilatory cycles solves the question of whether a chemoreflex drive exists or not, and gives an estimate of its relative importance (fig. 5.9). Chemoreceptors of this type were discovered by Heymans and his coworkers (see Heymans and Neil, 1958). In those mammals which have been studied, the glomi containing the chemoreceptors are perfused by arterial blood, and the receptors are excited not only by hypoxia, but also by hypercapnia, acidosis and certain drugs, the most commonly used in their investigation being NaCN. The receptor morphology is well known, but the mechanism of excitation, now the subject of much study, is as yet unsolved. Because there are other types of chemoreceptors affecting ventilation, those with the properties just described are sometimes called 'Heymans-type chemoreceptors'.

Chemoreceptors and chemoreflexes have been fairly completely investigated on very few species, mainly mammals, as table 5.2 shows, but they exist in all vertebrate classes. Those of birds are identical to the mammal's (see Bouverot, 1978). Chemoreceptors having exactly the properties of those of the mammal and bird have been reported in reptiles (two chelonians) and one amphibian, and although they are activated by stimuli which are not borne by purely arterialized blood, they should nevertheless be considered as Heymans-type chemoreceptors.

In the few aquatic water breathers which have been studied, the chemoreceptors appear to play an enormous role, since the ventilatory flow is very dependent on the oxygenation of the medium (fig. 5.1). But the existence of a chemoreflex has been demonstrated in very few species: the crayfish (Massabuau *et al.*, 1980) and maybe the lugworm (Toulmond and Tchernigovtzeff, 1984).

What is the role of hypoxia-induced ventilatory chemoreflexes? It has been shown that high-altitude hyperventilation is abolished if the chemoreflexes are suppressed by the ablation of the chemoreceptors or by section of their afferent connections with the respiratory centers (Bouverot, 1985). In an animal whose Heymans-type chemoreflexes have been suppressed, the tolerance to high altitude is markedly decreased. In human subjects born at high altitude *and* affected by chronic mountain sickness, the tolerance to hypoxia is very limited and it has been demonstrated that they do not have a hypoxia-induced chemoreflex drive (Lefrançois *et al.*, 1968).

TABLE 5.2

Some animals in which the evidence for Heymans-type chemoreceptors and chemoreflexes is sufficient. Morphological, physiological and electrophysiological evidence is complete in a few species only. The table is incomplete, especially for mammals. Heymans-type chemoreceptors may exist in the lugworm (Toulmond and Tchernigovtzeff, 1984).

	Morphological evidence	Physiological evidence	Electrophysiological evidence
Crayfish	–	14	–
Tench, carp, trout	–	8	–
Tuna	–	–	16
Toad	12	11	11
Tortoise	13	2	13
Hen	1, 3	5	5
Ox	1	4	–
Rabbit	1, 3	6	18
Cat	1, 3	15	9
Dog	1, 3	6	7
Man	1, 3	7, 10, 17	–

1 Adams (1958)
2 Benchetrit *et al.* (1977)
3 Biscoe (1971)
4 Bisgard and Vogel (1971)
5 Bouverot and Leitner (1972)
6 Bouverot *et al.* (1973)
7 Dejours (1962)
8 Eclancher and Dejours (1975)
9 Fitzgerald and Dehghani (1982)
10 Heymans and Neil (1958)
11 Ishii *et al.* (1966)
12 Ishii and Oosaki (1969)
13 Ishii *et al.* (1985)
14 Massabuau *et al.* (1980, 1984a)
15 Miller and Tenney (1975)
16 Milsom and Brill (1986)
17 Torrance (1968)
18 Verna *et al.* (1975)

It is not excluded that a decrease of blood oxygenation can act directly on the respiratory centers (pathway 3a of fig. 5.8), leading to a depression of the central respiratory activity. However, the mechanism of the central effect of blood oxygenation changes is unclear (Millhorn *et al.*, 1984; Bouverot, 1985).

Ambient carbon dioxide

The increase of ambient carboxygenation leads to an increase of ventilation. Possibly ambient hypercapnia can stimulate tegumental and mucosal receptors and entail a ventilatory response, counteracting a hypercapnia of the body fluids (fig. 5.8, pathway 2). However, any sizeable increase of ambient carboxygenation leads to a bodily hypercapnia which may stimulate various types of receptors (table 5.3). In probably all vertebrates, the respiratory centers and the Heymans-type chemoreceptors are sensitive to hypercapnia (pathways 3a and 3b). The mechanism of stimulation of the RC by hypercapnic and acid blood perfusing the brain stem has spawned an abundant literature (see Dempsey and Forster, 1982).

Whereas the central action of hypercapnia has been known since the end of the 19th century and the Heymans-type chemoreceptors and chemoreflexes since 1930, a new category of CO_2-receptors has been described during the last twenty years. These receptors are pulmonary (Milsom, 1979; Scheid *et al.*, 1978; Shelton *et al.*, 1986).

Some are chemoreceptors; their response is *inversely* proportional to P_{CO_2} and their activity inhibits the RC. Thus if P_{CO_2} increases, their discharge decreases, the RC are less inhibited and respiration increases. These receptors are not stimulated by mechanical deformation. Such CO_2-chemoreceptors exist in the few species of reptiles (Glass and Wood, 1983) and birds (Fedde and Kuhlmann, 1978) which have been studied.

On the other hand, mechanoreceptors of some amphibians and reptiles, certain of the classical pulmonary stretch receptors are also sensitive to low CO_2. For a given mechanical stimulation of these mechanoreceptors, hypercapnia decreases the afferent discharge; since by the mechanoreceptive activity of these receptors, the RC are inhibited (the classical Breuer–Hering reflex), it follows that hypercapnia decreases their inhibitory influence and, by this mechanism, can entail a hyperventilation. But this mechanism of hyperventilation may not play an important role in the range of normal variations of P_{CO_2} values.

These two categories of pulmonary receptors, the purely CO_2-chemoreceptors and CO_2-sensitive mechanoreceptors, are not stimulated by hypoxia or NaCN; they are not Heymans-type chemoreceptors.

Table 5.3 presents a summary of pathways by which CO_2 affects ventilation. Whereas in mammals most of the CO_2-induced hyperventilation is due to the stimulation of Heymans-type chemoreceptors and central sensors, in birds and reptiles the situation is more complex because of the pulmonary receptors inversely responsive to CO_2 increase, which undoubtedly respond in the normal range of variations of P_{CO_2}.

Furthermore, ventilation is a cyclic phenomenon during which P_{CO_2}

TABLE 5.3

Different types of CO_2-sensitive receptors in vertebrates except fishes. The evidence of their existence is often scanty, particularly in lower vertebrates, and concerns only one or two species of a given class. The physiological role of pulmonary CO_2-sensitive receptors is not clear.

	Effect of hypercapnia on ventilation of:			
	Amphibians	Reptiles	Birds	Mammals
Central chemoreceptors	?	+ (1, 1a)	+ (12)	+ (10)
Heymans-type chemoreceptors	+ (7)	+ (8)	+ (2, 3, 11)	+ (3, 6)
Intrapulmonary CO_2-chemoreceptors		+ (9)	+ (5)	
Pulmonary CO_2-sensitive mechanoreceptors	+ (5)	+ (5, 9)		+? (3, 4)

+ indicates an increase of ventilation.

1 Benchetrit and Dejours (1980)
1a Davies and Sexton (1987)
2 Bouverot and Leitner (1972)
3 Bouverot (review) (1978)
4 Coleridge *et al.* (1978)
5 Fedde and Kuhlmann (review) (1978)
6 Heymans and Neil (review) (1958)
7 Ishii *et al.* (1966)
8 Ishii *et al.* (1985)
9 Jones and Milsom (1979)
10 Kellogg (review) (1964)
11 Nye and Powell (1984)
12 Sébert (1979)

varies in the airway gas, the alveolar gas and even in the arterial blood. The discharges of the various CO_2-sensors are phased with the respiratory cycle. The ways in which these cyclic chemoreceptor discharges can constitute information to the RC and drive (or repress) ventilation is a much debated question (Cunningham, 1987).

Whatever the mechanism of the CO_2-sensitivity, the ventilatory response considerably limits bodily hypercapnia. Let us suppose that an air breather breathing air, with a value of arterial P_{CO_2} of 35 Torr, is switched onto a mixture containing 30 Torr of CO_2 (about 4% CO_2 at sea level). If there were no ventilatory response to CO_2, one should observe, after a very long time due to the readjustments of the CO_2-stores, an increase of 30 Torr in the body fluids; the arterial P_{CO_2} should tend to a value of 65 Torr, leading to a very marked acidosis which would cause grave disturbances in the function of many organs, particularly the brain. Bu the hypercapnia-induced hyperventilation increases the CO_2 clearance and limits the arterial hypercapnia, for example to 42 Torr, that is an increase of 7 Torr over the value observed when breathing normal air. If, in our example, the CO_2 pressure in the inspired gas is higher than 30 Torr, the ventilation is higher; if it is less than 30 Torr, the hyperventilation is less marked. One may consider that the increase of P_{CO_2} during CO_2 breathing leads to a hyperventilation which is not sufficient to return P_{CO_2} to its reference air-breathing value, which may be viewed as the set point of P_{CO_2} of the normal subject breathing normal air. This kind of regulation is called proportional; that is, the ventilatory response is driven by and is proportional to the increment of the arterial P_{CO_2} value.

CHAPTER 6

Responses to variations of energy needs

Summary

Three factors may considerably change the energy expenditure of animals: ambient temperature, muscular activity and feeding. Only the first two are discussed here.

Ambient temperature variations have different effects on poikilotherms and homeotherms. In poikilotherms, exposure to cold causes a decreased O_2 consumption, whereas in homeotherms it is increased. For the poikilothermic turtle taken as an example, the reported data (Kinney *et al.*, 1977) encompass the whole respiratory system. (1) The O_2 consumption increases with temperature; Q_{10} is about 3. (2) The ventilation increases less than the O_2 consumption. (3) Pa_{CO_2} consequently increases and pH decreases with temperature. The pH decrease approximates Rahn's rule. (4) The blood flow increases with temperature. (5) The pulmonary and systemic blood pressures do not change much, evidence for large decreases of the pulmonary and systemic resistances to blood flow. The regulatory mechanisms of these reactions to thermal changes are little known.

'*Muscular exercise*' covers several patterns of muscular contraction which must always be clearly stated and the environmental conditions must be specified. A disregard of these particularities leads to a misunderstanding of the reactions to exercise and their regulation. Since the various steps of the O_2- and CO_2 transfer between the milieu and the cells during prolonged dynamic exercise, here shortly reviewed, are disposed in series, they must be exactly coupled in any prolonged exercise.

The question of regulation of the respiratory system is first stated in general terms. The control of breathing in a prolonged exercise of normal humans, with sudden onset and stop, at sea level, in a temperate environment, is taken as an example. The neurohumoral theory of this control best explains the reactions mediated by neurogenic and blood-borne ventilatory stimuli whose characteristics are suggested. In the blood flow regulation during the same type of exercise, several factors, mechanical, neurogenic (central and reflex) and humoral, may come into play simultaneously to ensure the coupling between ventilation and blood flow.

A detailed general understanding of the control of respiration in water breathers and in homeothermic and poikilothermic air breathers, performing different types of exercise, in a variety of environmental conditions, is not yet possible, because of the paucity of data.

Three main factors increase energy expenditure and oxygen consumption: temperature variations, exercise and nutrition, as first demonstrated by Lavoisier. In birds and mammals, nutrition causes a moderate increase of oxygen consumption, but in other animals, the increase of the energy metabolism due to nutrition may be very important. Some lower vertebrates and many invertebrates may fast long periods, sometimes even several years, in a state of estivation, hibernation or wintering over. When they emerge from this torpor to active life, their oxygen consumption may be relatively enormous. However, the nature of this increase is complex, because exercise, for example in the search for food, and temperature also change. Here I will consider only two relatively simple cases of variations of respiratory intensity: thermal changes and exercise.

Temperature

The effect of an ambient temperature change on energy expenditure is quite complex. In birds and mammals, the fall of ambient temperature increases the energy metabolism and the respiratory intensity. Exposure to cold of small mammals may increase their O_2 consumption as much as exercise at neutral temperature (Pasquis *et al.*, 1970). A higher metabolism during cold exposure can be mediated by the increase of some hormones: epinephrine or thyroid hormones (non-shivering thermogenesis) or by shivering, which is a form of muscular exercise, or by both.

Effects of temperature variations on the respiratory system

Poikilotherms, lower vertebrates and all invertebrates, contrary to homeotherms, react to an increase of ambient temperature by a higher respiratory intensity. However, no simple law governs the change of oxygen consumption with ambient thermal change; in particular, the change of $\dot{M}_{O_2}$ may be time-dependent. If a fish is exposed to cold water, its first reaction may be a fall of respiratory intensity, but after some time the animal acclimates to this new environmental condition and its respiratory intensity returns to the value initially observed at a higher temperature (Prosser, 1986).

The effect of thermal change on oxygen consumption and external respiration has often been studied. Since, in this book, respiration is considered as a system encompassing the whole body, the example used describes the temperature-induced variations of several factors involved in the transport of oxygen from the surroundings to the tissues. A turtle, *Pseudemys floridana* (Kinney *et al.*, 1977), was observed at several temperatures between 12 and 38 °C. Oxygen consumption, ventilation, cardiac output, the blood pressures and the acid–base balance of the aortic blood were measured. The body mass of the turtles ranged from 0.5 to 4 kg. By multiple regression analysis, the equations of most variables studied as functions of temperature and body mass were calculated. The independent variable is the temperature. The oxygen consumption is related to body temperature by the equation:

$$\log \dot{M}_{O_2} \cdot B^{-1} = 0.047\,T' - 0.24 \log B - 2.67 \qquad (6.1)$$

in which $\dot{M}_{O_2} \cdot B^{-1}$ is the mass-specific oxygen consumption in mmol · $min^{-1} \cdot kg^{-1}$, T′ the body temperature in °C and B the body mass in kg. This equation shows that $\dot{M}_{O_2} \cdot B^{-1}$ decreases with the increase of body mass, a well-known phenomenon (see Calder, 1984; Schmidt-Nielsen, 1984), but questions of scaling will not be dealt with here. Equation (6.1) and all the following equations have been normalized for a 1 kg turtle. Equation (6.1) becomes:

$$\log \dot{M}_{O_2} = 0.047\,T' - 2.67 \qquad (6.2a)$$

hence $\dot{M}_{O_2} = 0.00214 \times 10^{0.047\,T'}$ (6.2b)

Table 6.1 gives the numerical values of various respiratory characteristics of the external respiration, the blood oxygen transport and the arterial blood pH and P_{CO_2} in one turtle. From these values, figs. 6.1–6.3 were drawn.

Figure 6.1 illustrates eq. (6.2b). The Q_{10} is constant at 2.95.

Figure 6.2 shows the relation between the ventilation, the $\dot{M}_{O_2}$-specific ventilation and the coefficient of O_2 extraction from the air as functions of

TABLE 6.1

Oxygen consumption, external respiration and blood circulation variables as functions of body temperature in a 1 kg turtle, *P. floridana* (Kinney *et al.*, 1977). Column 1, body temperature; 2, oxygen consumption.

Top panel. Columns 3–8, variables of external respiration. Column 3, ventilation; 4, $\dot{M}_{O_2}$-specific ventilation; 5, O_2 concentration in inspired air; 6, O_2 extraction coefficient from respired air; 7, respiratory frequency; 8, tidal volume; 9, pulmonary blood flow; 10, pulmonary ventilation/perfusion ratio; 11 and 12, respectively arterial pH and P_{CO_2}.

Bottom panel. Columns 13 through 22, variables concerning blood flow and the O_2 transfer by blood. Columns 13 and 14, systemic blood flow and $\dot{M}_{O_2}$-specific systemic blood flow; 15 and 16, O_2 concentration in the systemic arterial blood and O_2 extraction coefficient from systemic arterial blood; 17, cardiac frequency; 18, stroke volume; 19 and 20, pulmonary artery and systemic arterial blood pressure; 21 and 22, pulmonary and systemic resistances to blood flow.

1	2	3	4	5	6	7	8	9	10	11	12
T′	$\dot{M}_{O_2}$	$\dot{V}_E$	$\dot{V}_E \cdot \dot{M}_{O_2}^{-1}$	$C_{I_{O_2}}$	$E_{g_{O_2}}$	f_R	V_T	$\dot{V}b_L$	$\dot{V}_E \cdot \dot{V}b_L^{-1}$	pHa	Pa_{CO_2}
(°C)	$\left(\frac{mmol}{min}\right)$	$\left(\frac{L}{min}\right)$	$\left(\frac{L}{mmol}\right)$	$\left(\frac{mmol}{L}\right)$		(min)	(L)	$\left(\frac{L}{min}\right)$			(Torr)
12	0.008	0.011	1.40	8.82	0.08	0.9	0.012	0.009	1.15		
15	0.011	0.013	1.23	8.70	0.09	1.0	0.013	0.013	0.99		
20	0.019	0.019	1.00	8.50	0.12	1.4	0.013	0.024	0.78		
22.5	0.024	0.022	0.90	8.40	0.13	1.6	0.014	0.032	0.69	7.69	20.8
25	0.032	0.026	0.81	8.29	0.15	1.8	0.014	0.043	0.61		
30.7	0.059	0.038	0.64	8.03	0.19	2.5	0.015	0.082	0.46	7.59	29.7
35	0.094	0.051	0.54	7.82	0.24	3.2	0.016	0.135	0.38		
36.2	0.107	0.055	0.51	7.76	0.25	3.4	0.016	0.155	0.35	7.51	34.8
38	0.130	0.062	0.47	7.66	0.27	3.7	0.017	0.190	0.32		

1	2	13	14	15	16	17	18	19	20	21	22
T'	$\dot{M}_{O_2}$	$\dot{V}b_{sys}$	$\dot{V}b_{sys} \cdot \dot{M}_{O_2}^{-1}$	Ca_{O_2}	Eb_{O_2}	f_H	V_S	P_{PA}	P_{SA}	R_L	R_{sys}
(°C)	$\left(\frac{mmol}{min}\right)$	$\left(\frac{L}{min}\right)$	$\left(\frac{L}{mmol}\right)$	$\left(\frac{mmol}{L}\right)$		$\left(\frac{1}{min}\right)$	(L)	(Torr)	(Torr)	$\left(\frac{Torr \cdot min}{L}\right)$	
12	0.008	0.013	1.63			8.4	0.0015	33.4	43.1	3700	3720
15	0.011	0.018	1.68	3.97	0.15	10.7	0.0017	33.9	43.7	2670	2630
20	0.019	0.033	1.75			16.0	0.002	34.7	44.5	1560	1480
22.5	0.024	0.044	1.79	3.75	0.15	19.6	0.0022	35.1	44.9	1190	1110
25	0.032	0.059	1.83			24.0	0.0024	35.5	45.3	910	830
30.7	0.059	0.114	1.93	3.30	0.16	38.0	0.003	36.4	46.3	490	430
35	0.094	0.190	2.01			53.7	0.0035	37.1	47.0	310	260
36.2	0.107	0.218	2.03	2.77	0.18	59.2	0.0037	37.3	47.3	270	230
38	0.130	0.270	2.07			68.4	0.0039	37.6	47.6	220	190

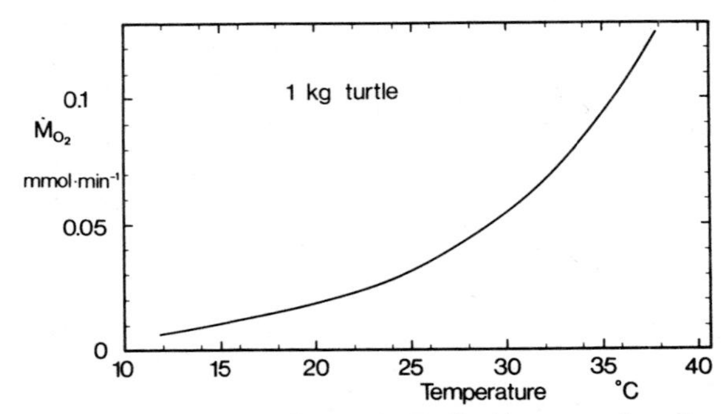

Fig. 6.1. Oxygen consumption in a 1 kg turtle *P. floridana* as a function of temperature (data derived from Kinney *et al.*, 1977).

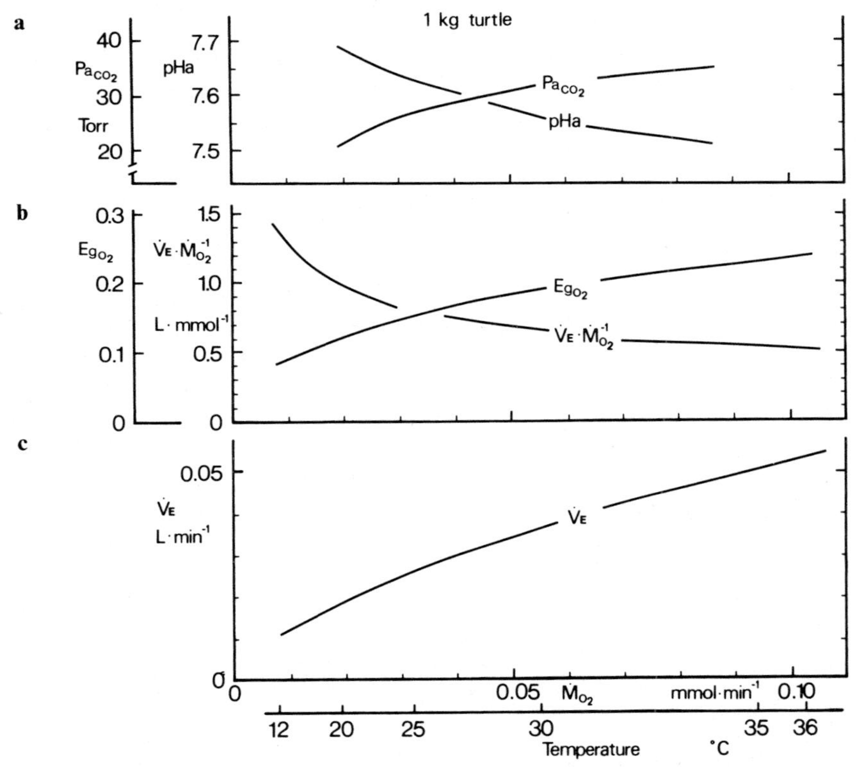

Fig. 6.2. As functions of oxygen consumption, $\dot{M}_{O_2}$, in a 1 kg turtle *P. floridana*: (a) aortic blood P_{CO_2}, Pa_{CO_2} and pH, pHa; (b) O_2 extraction coefficient from respired air, Eg_{O_2}, and $\dot{M}_{O_2}$-specific ventilation, $\dot{V}_E \cdot \dot{M}_{O_2}^{-1}$; (c) ventilation, $\dot{V}_E$. The abscissa scale $\dot{M}_{O_2}$ is doubled by a temperature line scaled according to eq. (6.1) and fig. 6.1 (data derived from Kinney *et al.*, 1977).

the oxygen consumption. The equations of these relations may be calculated from the data of Kinney *et al.* and eq. (6.2b). The ventilation increases as a function of $\dot{M}_{O_2}$, but to a lesser extent than the increase of $\dot{M}_{O_2}$; thus the specific ventilation, $\dot{V}E \cdot \dot{M}_{O_2}^{-1}$, decreases markedly and the coefficient of O_2 extraction increases. Similar trends were reported in *Pseudemys scripta elegans* by Jackson (1971) and Jackson *et al.* (1974).

The upper part of fig. 6.2 shows the changes of pH and P_{CO_2} in the systemic arterial blood. The values of pH fall with the increase of the oxygen consumption, *i.e.* with the increase of temperature, the mean fall of pH being 0.013 pH unit · °C^{-1}, which approximates Rahn's rule (see p. 36), whereas the arterial P_{CO_2} values increase. The $[HCO_3^-]$a remains nearly constant. These results of Kinney *et al.* (1977) in the turtle *Pseudemys floridana* confirm those of Jackson *et al.* (1974) in *P. scripta.*

Figure 6.3 shows some basic data for blood transport, namely $\dot{V}b_{sys}$, $\dot{V}b_{sys} \cdot \dot{M}_{O_2}^{-1}$ and Eb_{O_2}. The systemic blood flow is linearly related to, and almost directly proportional to, the oxygen consumption. Consequently, the ratio $\dot{V}b_{sys} \cdot \dot{M}_{O_2}^{-1}$ is almost constant and since Ca_{O_2} (table 6.1) does not vary much with temperature, the coefficient of O_2 extraction from blood is

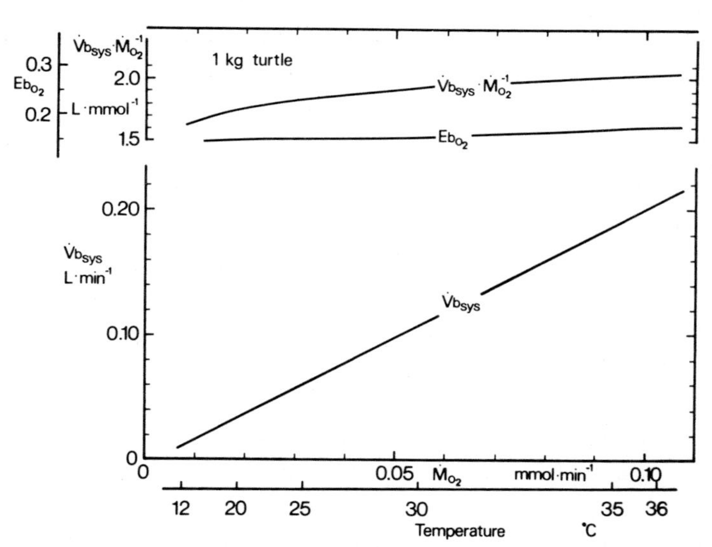

Fig. 6.3. As functions of oxygen consumption, $\dot{M}_{O_2}$, in a 1 kg turtle *P. floridana*: *Top*: $\dot{M}_{O_2}$-specific systemic blood flow, $\dot{V}b_{sys} \cdot \dot{M}_{O_2}^{-1}$ and O_2 extraction coefficient from systemic (aortic) blood, Eb_{O_2}. *Bottom*: systemic blood flow, $\dot{V}b_{sys}$. For abscissae, see fig. 6.2 (data derived from Kinney *et al.*, 1977).

nearly constant (see eq. 5.7). Kinney *et al.* (1977) reported also the pulmonary blood flow which is a little lower than the systemic blood flow (table 6.1, columns 19 and 20). The systemic and pulmonary blood pressures do not change with temperature, the mean values being respectively 31.7 and 25.1 Torr, with a constant difference of 6.6 Torr.

This remarkable set of phenomenological experimental data is very coherent and important because respiration is viewed at the organismic level. However, more data on the transfer of oxygen between the alveolar gas and the pulmonary capillary blood on the one hand, and between the systemic capillary blood and the cells on the other, are necessary to satisfy completely the requisites developed in the concept of a respiratory system (Rahn and Fenn, 1955; Dejours, 1966; Haab, 1982; Weibel, 1984; Otis, 1987).

Regulations

Very little is known about the *regulations* of the variations in the O_2- and CO_2 transfer system as a function of temperature, but certain points can be put forward.

(1) Tissue and organ respiration increases with temperature and the capillary beds may open by a local mechanism; indeed, columns 21 and 22 of table 6.1 show that the pulmonary and systemic resistances decrease with temperature.

(2) Blood viscosity decreases in bullfrogs and in painted turtles when the temperature is increased from 10 to 30 °C (Langille and Crisp, 1980); this phenomenon should facilitate the increase of blood flow.

(3) The modest increase of the stroke volume (column 18) may result from an increase of the venous return and the cardiac filling during diastole, according to Starling's law. Presumably the temperature increase raises the heart frequency (column 17) by a direct action on the cardiac pacemaker.

(4) It is not known whether the heart and the vascular smooth muscles are influenced by the change of activity of the autonomic nervous system as a result of the thermal change.

(5) The ventilatory activity increases as a result, presumably, of the direct action of temperature upon the brain and the respiratory centers. Kinney *et al.* (1977) report that the respiratory frequency, f_R, augments with temperature and also that V_T increases, but less markedly than f_R (columns 7 and 8).

(6) In the tortoise, *Testudo horsfieldi*, Benchetrit *et al.* (1977) demonstrated a ventilatory oxygen drive (chemoreflex drive) which is increased by a rise in temperature. The CO_2 drive of ventilation (Benchetrit and Dejours, 1980) may also be temperature-dependent. Chemoreflex drives presumably exist in the turtle *P. floridana* as well. Arterial P_{CO_2} increases with temperature (column 12), and since Kinney *et al.* report that P_{O_2} values decrease with temperature, these changes of P_{CO_2} and P_{O_2} may be factors of an increase of the chemoreflex drive of ventilation.

(7) There is mechanical interference between the circulation and ventilation, since the heart and blood vessels are located in the body cavity which also acts as an air pump. This disposition may be important in the necessary coupling between ventilatory and blood convections (p. 87).

(8) The results of Kinney *et al.* are mean values. For an animal whose breathing pattern 'consists of relatively brief periods of eupneic breathing followed by more protracted apneas' (*loc. cit.* p. 315), these mean values may not be the best ones, although it is hard to see how better ones could be obtained. Is there perhaps some special relation between the irregular ventilatory activity and the heart pumping? Jacob (1980) reported that in the seven species of snakes he studied, the cardiac frequency was higher during the ventilatory periods than during apnea. This observation is one of many examples (see following subchapter) of the interrelation between the cardiac and ventilatory pumps.

(9) In some poikilotherms exposed to an environment with a thermal gradient there may be interferences between the selected ambient and bodily temperature, the thermopreferendum, and the conditions of oxygenation. Hicks and Wood (1985) studied two species of *Varanus* and two species of *Iguana*. Exposure of the animals to hypoxia in a thermal gradient reduced the selected temperature, *i.e.* the animals chose an ambient temperature which is less energy-demanding. On the other hand, if by bleeding the blood hematocrit was reduced by 50%, the animals selected a lower ambient temperature until its blood O_2-carrying capacity returned to normal. These observations favor the concept that bodily reactions aim at ensuring a proper oxygenation.

Muscular exercise

There are many types of muscular activity. The following factors must be taken into consideration.

(1) Extent of body involvement: localized (typing), regional (walking) or global (cross-country skiing).

(2) Duration of the exercise: a short burst (jump, sprint); a few minutes' exercise; a prolonged exercise.

(3) Intermittent or continuous exercise.

(4) Posture: recumbent, seated, standing.

(5) Intensity of the exercise, from mild to strenous.

(6) Characteristics of the animal: species, age, size, degree of training.

(7) Environmental conditions of the exercise, the main ones being: the medium breathed – water or air; barometric pressure – hypobary (altitude), normobary (sea level), hyperbary (diving); gravity – terrestrial gravity, satellite microgravity; temperature and humidity – the main factors conditioning the heat loss. Some 80% of the energy used to perform a dynamic exercise is transformed into heat.

These factors create a variety of exercises which have never all been exhaustively considered, and will not be here. But every time one describes physiological reactions to exercises and their regulatory mechanisms, the type of muscular activity, in terms of the factors above, must be carefully stated.

Reactions of the respiratory system during prolonged exercise

Prolonged exercise, as running at sea level, may increase the respiratory exchanges several times, up to 20 times the resting rate in some mammals. However, the metabolic scope is quite variable with the species, the degree of training, the ambient temperature and humidity. In some mammals, particularly in humans, the respiratory reactions of the organism to exercise are very well known (Ekelund and Holmgren, 1964; Åstrand and Rodahl, 1977).

In water breathers, the results are scanty. Some basic data have been reported by Stevens and Randall (1967) who studied moderate swimming activity (four times the resting oxygen consumption) in the rainbow trout. More complete data have been gathered by Piiper *et al.* (1977) who studied a light swimming activity in the dogfish *Scyliorhinus stellaris*, whose $\dot{M}_{O_2}$ value during exercise was less than twice the resting value.

In steady state of exercise as compared to resting conditions, the flux of oxygen from the surroundings to the cells may be analyzed in terms of conductance according to the general equation

$$\dot{M}1,2_{O_2} = G1,2 \times (P1_{O_2} - P2_{O_2}) \qquad (6.3)$$

in which 1 and 2 designate two locations in the respiratory system:

Step 1: ambient medium to the branchial or pulmonary medium;

Step 2: branchial or pulmonary medium to the postbranchial or postpulmonary blood (arterialized blood);

Step 3: arterial blood to venous blood;

Step 4: capillary blood to body cells.

This approach was used in Dejours (1981, pp. 130–133) who analyzed the data of Ekelund and Holmgren (1964) on exercising humans with an oxygen consumption near 9 times the resting value, and in Taylor and Weibel (1981) and Weibel (1984) who studied various mammals of different size from 7 g pygmy mice to 105 kg ponies. Since the animals were studied in nearly steady state at rest or in exercise, all the values of the net O_2 flux between two locations of the respiratory system are almost equal.

Step 1: ambient medium to the branchial or pulmonary medium. Oxygen conductance increases, due to an increase of the ventilatory flow. This increase is roughly in direct proportion to the increase of oxygen consumption. Consequently the branchial mean P_{O_2}, taken as the midpoint between inspired and expired water P_{O_2} values, and the pulmonary (alveolar) mean P_{O_2} are about constant. However, in humans performing strenuous exercise, the increase of ventilation may be considerably higher than the increase of oxygen consumption; alveolar hyperoxia and hypocapnia ensue. In prolonged exercise in hot and humid ambient conditions, unfavorable to heat loss, animals like dogs which do not sweat use ventilatory maneuvers to dissipate heat and pant. Then if the heat stress is strong enough, relative alveolar hyperventilation may occur with hypocapnia (see p. 102).

Step 2: branchial or pulmonary medium to the postbranchial or postpulmonary blood (arterial blood). The conductance of this step is complex, as explained in all books of respiratory physiology. But at rest, the local diffusing capacity has a potential reserve which may be increased during exercise in water breathers and air breathers, so that the arterial P_{O_2} values do not decrease markedly, at least in normoxia.

Step 3: arterial blood to venous blood. The higher conductance is due to (1) an increase of blood flow, (2) an increase of the blood O_2 capacitance. The lesser venous O_2 saturation along the steeper part of the O_2 dissociation curve increases the slope of the line joining the arterial point to the venous point, *i.e.* the arterial–venous O_2 capacitance. The Bohr effect and, eventually, an increase of blood temperature in the exercising muscles are factors which augment the arterial–venous blood O_2 capacitance.

Step 4: capillary blood to body cells. In the exercising muscles, the blood flow increases: the terminal arterioles to the muscles open, and more capillaries are patent (Krogh, 1922; Lindbom, 1983). The advantage of the functional hyperemia of exercising muscles is that the distance between the capillary blood and the muscular cells decreases. These phenomena have been demonstrated in mammals, and although they may be general, Stevens (1968) observed no obvious indication of their existence in exercising trout.

The increase of the muscular blood flow implies an increase of the cardiac output, although some organs like the viscera, except for the heart, may be underperfused for some time. In any case the cerebral circulation is maintained.

The carbon dioxide efflux and the oxygen influx have the same order of magnitude. The increase of the various conductances favors the transfer of CO_2 as well as that of O_2. The net result is that the arterial blood P_{O_2}, P_{CO_2} and pH in fish and upper vertebrates are about the same in exercise and at rest. This does not mean there are no changes in the P_{O_2}, P_{CO_2} and $[H^+]$ values; there are usually some which depend in particular on the intensity of the exercise and on the animal species.

These reactions to prolonged steady-state exercise are not those of a very short exercise, as a jump or a brief movement of escape or attack. Then the oxygen consumption is not immediately increased, since the energy necessary for a short bout of muscular activity is drawn from the high-energy phosphate reserves in the muscle. In this case there occurs a depletion of the phosphagens, which can be replenished later on by aerobic metabolism over a protracted time period. The rate of increase in the four steps previously mentioned may then be modest and difficult to observe.

Regulations

Figure 5.8 (p. 63) shows that there are three categories of stimuli of the respiratory centers: intercentral (pathway 1); reflex pathways 2 and 4; blood-borne stimuli (pathway 3). Zuntz and Geppert (1886) understood very well that they could be involved in the regulation of breathing in exercise.

Since Zuntz and Geppert, thousands of articles have been devoted to this problem. Dejours (1959, 1964) gave a nearly complete review of the past and then-current theories and reviews by Cunningham (1974), Whipp (1981) and Asmussen (1983) summarize further studies.

In my own reviews, I classified the theories as follows:

(A) Exclusively humoral theories: (1) humoral factors of unknown nature; (2) carbon dioxide; (3) hydrogen ion; (4) several specific humoral factors.

(B) Exclusively neurogenic; either intercentral, or reflex, or both.

(C) Neurohumoral.

New views or explanations proposed enter necessarily into one of the three categories. I consider a neurohumoral theory as the most realistic. When one looks at the list of stimuli (fig. 5.8) which may be changed during exercise, it appears highly improbable that one single factor or a single group of factors is responsible for the ventilatory changes of exercise. There is in fact much direct evidence that *several* factors come into play in the course of exercise.

The neurohumoral theory takes into account various types of exercise, dynamic or static; at sea level and at high altitude; in temperate, cold or hot climates. Concerning the complexity of the problem, 'The considerable variety of types of exercise result in a number of physiological states having muscular contraction as their only point in common. The pattern of ventilatory regulation, while stemming from a certain grouping of neurogenic and humoral factors, is also variable and cannot be generalized for all types of exercise' (Dejours, 1959, p. 249).

Although the concept of respiration, with the various steps between the cells and the environment was clearly recognized in 1959, our knowledge of the regulation of circulation was insufficient to propose a theory encompassing the regulation of all the elements of the respiratory system. 'It is evident that the study of the ventilation which ensures the O_2 and CO_2 exchanges between the alveolar atmosphere and the ambient atmosphere cannot be separated from the study of the gas exchanges between the alveolar gas and the tissues, which is a problem of circulation. The interminglings between the ventilatory and circulatory apparatus are multiple anatomically and functionally as well as by their control mechanisms; the regulations of these apparatuses work always *pari passu* and it is artificially and temporarily that one has dealt until now with the separated study of each of them' (Dejours, 1959, p. 246).

An example of control of breathing in exercise: prolonged exercise, with abrupt start and stop, at sea level, in temperate environment, in a normal human subject

Figures 6.4 and 6.5 show the change of ventilation at the start and stop of exercise, in which there is no inertia to overcome at the start and no deceleration at the end of the exercise. Figure 6.5 shows the following sequence of phases: (1) rest; (2) abrupt rise of $\dot{V}_E$ at the onset of exercise; (3) an 'on' plateau lasting 20–40 sec; (4) a several-minute period of exponential increase of ventilation terminating in the ventilatory steady state (5); (6) a sharp fall at the cessation of exercise; the fall is sometimes progressive and lasts 2 or 3 cycles, but the time constant does not exceed a few seconds; (7) a plateau lasting 20–40 sec, a duration which may differ from the 'on' plateau; (8) a slow decrease of ventilation during the recovery period which is uniexponential in mild exercise, and biexponential in heavy exercise.

These observations are old; in particular the abrupt change of breathing at the onset of exercise was described by Krogh and Lindhard in 1913. However, these authors (1919) did not detect the abrupt ventilatory change at the transition from work to rest which nevertheless is visible in their figure. This kind of curve and its interpretation was systematically studied by Dejours *et al.* (1955); the description of the curve was adopted by Wasserman *et al.* (1986). The fast changes seem to be neurogenic, either centrogenic, or reflex, or *both* (see Dejours, 1959, 1964; Fordyce *et al.*, 1982). The slow changes of ventilation (phases 4 and 8 of fig. 6.5) are more difficult to explain. It is possible that the local physicochemical changes which occur during the exercise may stimulate peripheral receptors in the moving parts of the body. A centrogenic stimulus of ventilation may also increase during exercise, especially when the subject finds it difficult to prolong the exercise. However, some chemical blood-borne factors come into play, because phase 8 of fig. 6.5 is

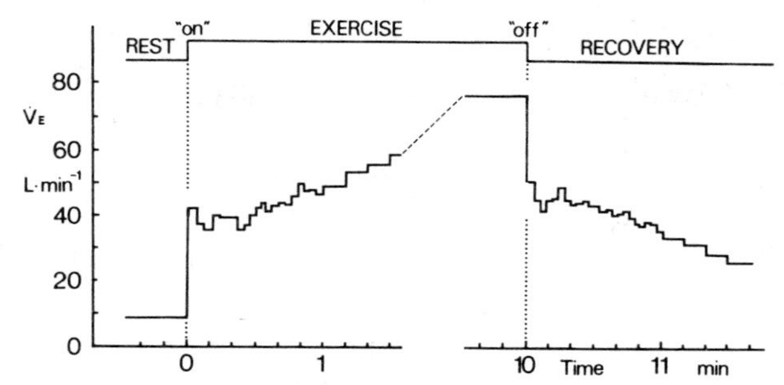

Fig. 6.4. Ventilatory flow rate, V̇E, at rest, during an exercise on a Fleisch bicycle at 60 revolutions/min and 180 W, and during recovery. The onset (without inertia to overcome) and the cessation of the exercise are instantaneous. During the first minute of exercise and the first minute of recovery each bar corresponds to the value of V̇E during one breath. Thereafter each bar is the mean ventilation for a 10 sec period. The ventilation from the end of the second minute until the last 30 sec of the 10 min exercise is skipped (from Dejours and Teillac, 1963).

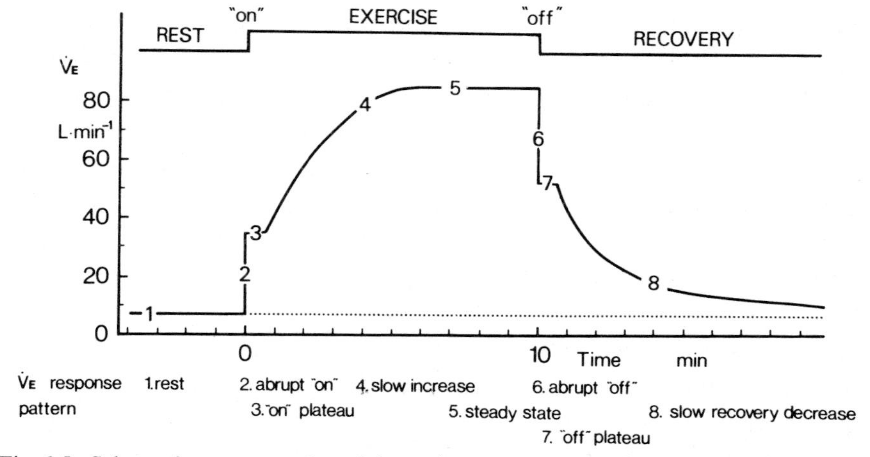

Fig. 6.5. Schematic representation of the various components of ventilation at rest, during prolonged exercise and during recovery (Dejours *et al.*, 1955). The exercise is of the type described in the legend of fig. 6.4.

suppressed when just before the end of a leg exercise the blood flow through the legs is stopped by suddenly inflating cuffs placed around the thighs.

This is the essential of the neurohumoral theory of the control of breathing in this type of exercise, as presented in the late fifties.

Nature of the neurogenic stimuli. There is ample evidence that:

(1) the brain activity during exercise, in particular the hypothalamus, can provide simultaneous drives for locomotion, and for respiratory and circulatory changes (Eldridge *et al.*, 1985). Furthermore, there exists some plase-locking of the frequency of movements and the ventilatory frequency in mammals and birds (Bramble and Carrier, 1983);

(2) stimuli originating in the moving parts of the body reflexly induce respiratory and circulatory changes (see Dejours, 1964). The nature of the limb receptors involved is the subject of several studies (see Whipp, 1981);

(3) as will be discussed below, at the onset of exercise venous return and pulmonary blood flow immediately increase. The alternative contractions of agonist and antagonist muscles pump blood in and out of the muscles, an important factor of the increased venous return. Possibly, there is some information originating from the thoracic blood vessels or from the heart baro- or stretch-receptors eliciting breathing reflexes (see Dejours, 1959, 1964; Whipp, 1981).

Nature of the blood-borne stimuli. Humans at sea level display a chemoreflex drive of breathing, and also a CO_2-H^+ drive, which at least partially originates in the arterial chemoreceptors (Heymans-type receptors). The catecholamines which are considerably increased above the anaerobic threshold may contribute to the ventilatory drive (Flandrois *et al.*, 1977; Whipp, 1981). Other humoral factors have been invoked. Recently it has been shown that plasma $[K^+]$ increases during exercise and evidence has been given that hyperkaliemia may increase ventilation via a chemoreflex mechanism (Band *et al.*, 1985). The concentration of the conjugate pair NH_3/NH_4^+ is increased in heavy exercise (Dudley *et al.*, 1983) and it has been shown that its artificial increase by infusion results in hyperventilation (Wichser and Koizumi, 1974).

In the Pekin duck, and presumably in all birds, a special blood-borne drive of breathing may exist. As seen on p. 67, birds possess CO_2-sensitive intrapulmonary chemoreceptors (IPC) whose change of stimulation by hypercapnia increases the ventilation. Hypercapnia and acidosis have been observed in the mixed venous blood of ducks performing a treadmill exercise which tripled the O_2 consumption (Kiley *et al.*, 1985). Since this mixed venous blood reaches the IPC's (see Bouverot, 1985), it follows that the variations of their stimulation during exercise may entail a blood-borne reflex control of breathing. Mixed venous blood chemoreceptors have not been demonstrated in mammals (see Dejours, 1964).

Besides the absolute values of arterial P_{O_2}, P_{CO_2} and pH, whose deviations, in humans, from reference values or set points may constitute respiratory stimuli, the oscillations of these factors, locked in with the respiratory cycles, are thought to be potential, and possibly very important, ventilatory drives. Whipp (1981, pp. 1110–1114) gives a detailed account of how these P_{CO_2} and pH oscillations may stimulate breathing.

In dogs (Favier *et al.*, 1983), quail (Nomoto *et al.*, 1983), and the dogfish *Scyliorhinus stellaris* (Piiper *et al.*, 1977), whose ventilation has been continuously recorded, it is clear that there is an abrupt 'on' increase and an 'off' decrease of ventilation at the start and the cessation of exercise, and that the schematic curve presented in fig. 6.5 holds true for these animals. Ventilation in the fish may be mechanically increased by swimming without special ventilatory adaptation; but there is anyway some neurogenic control since the ventilatory frequency shows 'on' and 'off' responses (Piiper *et al.*, 1977).

The control of ventilation and blood flow

The question of regulation concerns the whole respiratory system at rest and during exercise. It is known that in humans active motion immediately provokes not only an increase of ventilation, but also an increase of the O_2 uptake by the blood flowing through the pulmonary capillaries (Krogh and Lindhard, 1913; Dejours *et al.*, 1956; Raynaud *et al.*, 1973; Beaver *et al.*, 1981; Cerretelli and Di Prampero, 1987). The increased O_2 uptake by blood means that there is an increase of cardiac output. There is furthermore a concomitant increase of the heart frequency, which implies some immediate change of the autonomic control of the cardiac pacemaker (Dejours *et al.*, 1956; Raynaud *et al.*, 1973; Duffin and Bechbache, 1983). Rapid changes of the pulmonary blood flow at the beginning and at the end of exercise have also been reported in humans by Cerretelli *et al.* (1966), and are implied by the observations made by Beaver *et al.* (1973). The sequence of events presented in fig. 6.5 for ventilation is also true for cardiac output and oxygen consumption, even if there may be some quantitative differences, due in particular to the fact that the arterio-venous O_2 and CO_2 concentration differences vary markedly in exercise, whereas the inspired–expired or inspired–alveolar changes are less pronounced.

The coupling of circulation and ventilation in the concept of gas exchange system of the body is a necessity for explaining the O_2 delivery to and the CO_2 clearance from the cells. The mechanisms of the coupling raise several problems, but the following factors must be considered.

(1) Mechanical interplay: the cardiac pump is located within the gas pump, the 'thorax', in reptiles, birds and mammals. In humans, voluntary hyperventilation, mainly that resulting from an augmented tidal volume, facilitates the return of the venous blood to the thorax and increases the systolic volume and the cardiac output; the increase of blood flow is more marked in the sitting than in the supine position. The circulatory effect of voluntary hyperventilation is purely mechanical, since no change in heart frequency is observed (Boutellier and Farhi, 1986).

(2) Hypothalamic stimulation leads to concomitant locomotion, increase of breathing and increase of circulation (Eldridge *et al.*, 1985; Stone *et al.*, 1985).

(3) Some stimuli originate in the moving parts of the body and lead to parallel ventilatory and circulatory responses, including change of cardiac frequency (Dejours *et al.*, 1956; Stone *et al.*, 1985).

(4) The changes of some blood-borne factors, as catecholamines, lead to ventilatory and circulatory responses.

Obviously, the regulation of ventilation and circulation is quite complex, but it is clear that one has to deal with both functions together and that several factors interplay normally.

Most studies on the regulation of the gas exchange system in exercise have been performed on human beings, a few on dogs (see Szlyck *et al.*, 1981 and Favier *et al.*, 1983), some on cats (see Eldridge *et al.*, 1985) and on ponies (Forster *et al.*, 1984; Powers *et al.*, 1987), very few on birds (Kiley and Fedde, 1983; Nomoto *et al.*, 1983; Faraci *et al.*, 1984) and lower vertebrates (Sutterlin, 1969; Heath, 1973; Mitchell *et al.*, 1981).

In invertebrates, the control of breathing during muscular activity is little investigated. Herreid *et al.* (1983) have studied the respiratory and cardiac reactions of land crabs, *Gecarcinus lateralis*, running on a miniature treadmill. The authors did not report 'on' and 'off' transients of ventilation and the heart, and the crab's ventilatory and circulatory responses are so slow during exercise and during recovery that it seems certain that the pattern of responses in these animals cannot follow the description of fig. 6.5. Actually the increase of oxygen consumption is progressive, and a protracted period of recovery towards the pre-exercise values is constantly observed.

In spite of the scarcity of data regarding ventilation and circulation during exercise in water-breathing invertebrates, several works in arthropods provide beautiful examples of circulatory and ventilatory coupling. For example, in the crayfish *Astacus leptodactylus* (Angersbach and Decker, 1978), and the crabs *Cancer productus* (McMahon and Wilkens, 1977) and *Cancer pagurus* (Bradford and Taylor, 1982), the activity of the scaphognathites and the heart is intermittent. Synchronous periods of scaphognathite beating and heart beating are interspaced by periods of complete inactivity, or decreased activities. In these species the heart beat is neurogenic; presumably some neural control coordinates the beating of the heart and of the scaphognathites, then allowing simultaneous ventilatory and circulatory regulations (Taylor, 1982; Wilkens, 1982). But, in the amphibian tadpoles whose heart beat is primarily myogenic, cardio-ventilatory synchrony is commonly observed (Wassersug *et al.*, 1981).

The overview given in this chapter is marred by many shortcomings. In the respiratory system, it is difficult to designate the controlling variable, the control system, and the controlled signal, as defined in fig. 2.1. Each of the elements of this complex regulation has to be discussed.

Question 1: Which blood: arterial, capillary, or venous, is controlled? It is probably the arterial blood; however, the arterial blood is the same everywhere and the organs have different needs as to O_2 delivery and CO_2 and proton clearances. Since the arterial blood is the same at the entry of all organs, the only possible way for a given organ to have the most favorable local conditions is to regulate its own blood flow. For the organs themselves, the controlled signals are presumably the status of O_2, CO_2 and protons in their own capillaries or in their neighborhood.

Question 2: What becomes of the set point for CO_2 and CO_2-H^+ status during exercise as compared to rest? We will limit the discussion to the air breathers because, in water breathers, the O_2 drive is predominant, the CO_2/H^+ drive subordinate or negligible and because, in any case, the problem in terms of control theory has not been raised for water breathers. For air breathers at rest, it is probably safe to state that the O_2 and CO_2/H^+ status exert proportional control upon the ventilation. This has been demonstrated for the CO_2/H^+ status in some conditions; it is less certain, although probably true, for O_2 at sea level. During exercise, that problem is more complex. Even at sea level there is a hypoxic drive of ventilation. But it is not certain at all that there is a CO_2/H^+ drive, because the concentration of CO_2 and H^+ may actually decrease in mild and moderate exercise in mammals (Dempsey *et al.*, 1985). Thus P_{CO_2} and pH of arterial blood in some animals are displaced in the wrong direction to be ventilatory drives. Perhaps there is a change of set point for P_{CO_2}/pH, but then the question is, what shifted the set point?

Question 3: Can the O_2, CO_2 and H^+ status in the blood be simultaneously regulated in all conditions? The answer in sea level residents is yes. But in acute hypoxia, at rest or during exercise, the ventilation increases because of the intense O_2 drive, so that P_{CO_2} and H^+ concentrations decrease; this is one of the reasons for considering that the satisfactory oxygenation of the blood takes precedence over proper ABB regulation. In chronic hypoxia, the pH may return to its sea level value in spite of a marked hypocapnia; this is possible by a fall of the [HCO_3^-]. In passing, this shows that the ultimate controlled signal in the $CO_2/HCO_3^-/H^+$ complex is the [H^+] value. When the hypocapnic alkalosis is compensated in high altitude residents, then the blood O_2 and CO_2/H^+ status can be simultaneously regulated as at sea level, but with a chemoreflex hypoxic drive much more intense than at sea level.

Question 4: Are the ventilation and the blood flow really the controlling signals? No, since these signals are each the result of two products: breathing frequency and breath volume for ventilation; heart frequency and stroke volume for the cardiac output. To consider only the ventilation or the blood flow as the controlling signals can be erroneous. For example, it may happen that at the end of a prolonged exercise in a normal human, there is no abrupt change of ventilatory flow, but an abrupt change of breathing pattern, consisting of an increase of tidal volume and a decrease of respiratory frequency.

Here I have dealt mainly with the ventilatory responses, a little with the cardiac response, very little with peripheral circulation, so that I am far from having given the whole picture. Furthermore, exercise taxes all the bodily functions, endocrine, renal, thermoregulatory, hepatic, *etc.*, and enhances the activity of the two systems which ensure the unity of the organism, namely the nervous system, autonomic and cerebrospinal branches, afferent and efferent pathways, and the circulatory system: in exercise efferent and afferent neural traffic is intensified and circulatory delays are shortened.

To understand the regulations of the physiological responses to the changes in the respiratory intensity, only an organismic approach is valid, and one must consider the various groups of animals, the different types of exercise, in all environmental conditions compatible with life.

CHAPTER 7

Interferences between respiration and other functions

Summary

Several functions of the organism and many variations of certain environmental factors may interfere with or disturb the work of the respiratory system, transitorily or more or less permanently. This kind of question is often overlooked. The respiratory system with its two main components, external respiration and blood circulation, carry out functions which have nothing to do with the O_2- and CO_2 transfers, and reciprocal interferences may exist between respiration and other functions (table 7.1). The extent to which the oxygenation and the carbon dioxide clearance of the body are interfered with is examined in some selected examples, taken among water breathers and air breathers.

Ventilatory irregularities or arrests

Each time external breathing stops, blood P_{O_2} decreases and P_{CO_2} increases. The combination is by definition an asphyxia.

Feeding

In all activities related to feeding: sucking, drinking, eating, swallowing, rumination, *etc.*, the periods of apnea are generally short and the asphyxia moderate.

A special case of interference between feeding and ventilation is observed in filter feeders. These animals, sponges, lamellibranchs and ascidians, circulate water through their body to obtain food as well as oxygen, and they trap microscopic plankton. Hazelhoff (1939) observed that the O_2 extraction is lower in these animals than in aquatic non-filter-feeders.

TABLE 7.1
Factors which affect respiration.

Environmental factors	Organismal factors
– Ambient conditions for heat dissipation: radiations, convection, temperature, humidity	– Feeding
– Vibration	– Vocalization
– Noise	– Attention, emotion, crying, laughing, purring, sighing, sniffing
– Ionic environment	– Hiccup
– Water and food availability	– Yawning
– Hydrostatic pressure	– Sleep
– Barometric pressure	– Physical effort
– Ambient oxygenation	– Exercise
– Gravity, acceleration	– Posture
	– Diving

It may be that the density of food particles in the water is so low that the rate of water flow required to ensure adequate nutrition is higher than that necessary to satisfy the respiratory needs of the animals. This simple reasoning may not be sufficient. A more complex explanation suggests that the rate of water filtered depends on the oxygen *and* food concentrations in the environment (Jørgensen, 1975).

Diving

With the exception of the rare animals whose lung is built for air- and water breathing, as the pulmonate snail *Lymnaea*, apnea is a characteristic of diving air breathers. Most animals do not dive very long, but some sometimes do, either to catch food or to escape from predators. It has been shown that imposed artificial dives of birds, such as ducks, or mammals, such as the harbor seal, may provoke circulatory reactions which are quite different from those observed in spontaneous natural dives. However, in certain diving mammals, as the Weddell seal (Kooyman, 1981; Kooyman *et al.*, 1981; Zapol, 1987), there are some profound physiological changes as originally described in 1940 by Scholander (see Scholander, 1964): a decrease of cardiac output with a preferential distribution of blood to the brain, a remarkable cerebral resistance to asphyxia, and the capability for organs other than the brain and heart to turn to anaerobic energy production and to actually decrease their energy requirement considerably, a reaction which implies a local fall of temperature.

Speech

Vocalization disturbs regular rhythmic ventilation. When a resting subject is obliged to speak loudly because of a high surrounding noise level, as may happen during meetings and in classrooms, he hyperventilates and becomes hypocapnic. When he stops talking, the hypocapnia entails a hypoventilation which lasts for some time until the alveolar CO_2 tensions and the CO_2 reserves are restored (Dejours *et al.*, 1967).

In exercise, above all strenuous exercise, speaking is difficult; ventilation decreases during the period of speech, and is followed by a period of stronger hyperventilation (Doust and Patrick, 1981). We know by experience that speaking is particularly difficult during climbing at high altitude, but I do not know of any systematic study.

Purring

A special case of a perturbed regular breathing pattern is the purring observed in some felines and studied in cats (Remmers and Gautier, 1972). Purring resulting from the intermittent activation of intrinsic laryngeal muscles at a frequency of 20–30 Hz is superimposed on the basic breathing pattern which is altered. The minute ventilation is increased by 70%, due to an increase of the respiratory frequency with no change in tidal volume. There is an alveolar hyperventilation leading to a hypocapnia (P_{CO_2} is approximately 26 Torr during purring against 32 Torr without purring).

Sleep

Ventilation and circulation are much affected by sleep. Most of the studies have been done in humans and cats. During slow-wave sleep, respiration is generally regular and alveolar P_{CO_2} and P_{O_2} are respectively higher and lower than in the awake state, but periodic breathing may be observed. In paradoxical (rem) sleep, the frequency of breathing may increase, and periodic breathing, which may take the form of the Cheyne–Stokes ventilation, is frequent. In the apneic phases, which may last several tens of seconds, circulation (cardiac output, cardiac frequency, arterial blood pressure) may then be also considerably altered. A long apnea entails a marked asphyxia. Apneic states may be observed at any age during sleep, but they are most common, alarming and threatening in premature babies. Within several weeks of a premature birth, the frequency of these apneic states usually decreases. There is an abundant literature concerning sleep and breathing (see Saunders and Sullivan, 1984; Krieger, 1985).

Posture, gravity, immersion

With changes of posture (lying, standing), gravity (normal earth gravity, the increased gravity of space flight at launch and at landing; orbital microgravity), or water immersion and emersion, two main mechanical changes occur: (1) the shape of the thoraco-abdominal cavity and the lung volume; (2) the distribution of blood in the body. These changes affect the external respiration and the circulation.

When an animal moves from recumbency to the erect posture, the blood volume in the lower part of the body increases and the venous return is hindered; the cardiac output decreases mainly because the stroke volume falls considerably, since concomitantly the heart frequency increases. The total and alveolar ventilations increase out of proportion to the small rise of the respiratory exchanges, so that alveolar P_{O_2} increases and alveolar P_{CO_2} decreases (Bjurstedt *et al.*, 1962). The ventilatory changes are presumably related to the impaired venous return, since the hyperventilation in the erect position is almost completely reversed when the hydrostatic effects of the postural changes are removed by water immersion (Anthonisen *et al.*, 1965). Nevertheless, the mechanism of all the reactions observed between the two positions appears quite complex (Farhi and Linnarsson, 1977).

On the other hand, in resting or exercising subjects exposed to an acceleration along a longitudinal axis of 2 or 3 times the ground gravity, such that blood is pooled in the lower part of the body, the changes of the circulatory and respiratory variables described above are augmented (Bjurstedt *et al.*, 1968; Boutellier *et al.*, 1985).

During the launch of a manned satellite, the passengers undergo a few minutes of increased gravity; several days or weeks of 'weightlessness' (microgravity) follow; during the flight back to earth hypergravity again occurs before the landing and the return to ground gravity. These changes of gravity entail marked changes in blood circulation. In particular, in orbital space flight compared to ground level conditions, there is a considerable headward shift of blood from the lower part of the body which has many consequences as to heart volume, blood pressure, blood volume and diuresis. There is some acclimation to prolonged microgravity which is called 'cardiovascular deconditioning' (Levy and Talbot, 1983). Indeed, when the subjects return to earth and the natural 1 G environment, they exhibit cardiovascular troubles. When they stand up, they may faint, presumably because, being 'deconditioned', their venous return is impaired, their arterial blood pressure falls and their cerebral blood flow is decreased. Little is known about pulmonary function itself in orbit. We may hope that future space experiments will provide many new facts regarding cardiovascular and respiratory functions in weightlessness (West, 1986).

Ionoregulation and respiratory regulation

In water breathers, the skin and gills exchange not only O_2 and CO_2 molecules, but also ions and water. It is thus not surprising that variations in the ionic composition of the water change the CO_2 status and the acid–base balance of the body fluids, since CO_2 is equilibrated with two ion species, HCO_3^- and CO_3^{2-}.

Euryhaline marine species may live in brackish water. For example, Truchot (1981) observed a metabolic alkalosis (0.2 pH unit) in the hemolymph of the green crab, *Carcinus maenas*, exposed to diluted seawater. Analogous findings were made in the euryhaline crayfish, *Pacifastacus leniusculus*, moved between 75% seawater and freshwater (Wheatly and McMahon, 1982).

Probably any change in the ionic composition of the water, if broad enough, can change the blood ABB of aquatic animals. The effect of variations of water $[Cl^-]$ on hemolymph ABB has been studied by Dejours *et al.* (1982) and Burtin *et al.* (1986) in crayfish. In all cases an increase of ambient $[Cl^-]$ results in an acidosis, and a decrease in an alkalosis. The ABB disturbance is mainly metabolic when the change in $[Cl^-]$ is balanced by a change in $[SO_4^{2-}]$ (Dejours *et al.*, 1982). But if the change of $[Cl^-]$ is accompanied by an identical change of $[Na^+]$, the change in ABB is nearly compensated by an appropriate change of ventilation and the blood P_{CO_2} (Burtin *et al.*, 1986).

The value of titration alkalinity, TA, of the ambient water may also affect ABB and may play an important role in buffering the carbon dioxide produced. If the water has a low pH and a high TA, the buffering of $CO_2(H_2CO_3)$ is marked since CO_3^{2-} is transformed into HCO_3^-. In this case, the relation between ambient C_{CO_2} and P_{CO_2} (fig. 3.2, p. 26) is curvilinear, that is, the water's CO_2 capacitance at low P_{CO_2} is high. However, if TA is low, or if the water P_{CO_2} is high (and the water pH low), no buffer is available and the CO_2 capacitance coefficient is close to the CO_2 solubility coefficient.

When the CO_2 capacitance is high, the blood P_{CO_2} differs little from the expired water P_{CO_2} and is in any case much lower than the value predicted by Rahn (1966) for unbuffered waters. On the other hand, with the decrease of TA and the buffering capacity of the water by addition of a strong acid, the freshwater crayfish *Astacus leptodactylus* (Dejours and Armand, 1980) and the marine crab *Carcinus maenas* (Truchot, 1984) develop a hypercapnic acidosis.

Figure 7.1 shows the considerable changes of the acid–base balance, ABB, of the hemolymph of the crayfish *A. leptodactylus* under the influence of *various environmental factors* (Dejours and Armand, 1982), grouped in such a way that they lead to an increase or a decrease of pH. The symbol R is the reference value of the hemolymph ABB for animals living in the reference water as defined by table 7.2. Then, step by step, the ambient chloride concentration was changed, a decrease in A, an increase in A′. A week later the water acid–base balance was then altered, to hypocapnic and metabolic alkalosis in AB; the water A′ to hypercapnic and metabolic acidosis in A′B′. Finally, a few days later, the ambient O_2 pressure, in its turn, was changed, to hypoxia in ABC, and to hyperoxia in A′B′C′. The whole experiment lasted 40 days at 13 °C.

In the situation ABC, the animal lived in a low chloride, hypocapnic and hypoxic water with a high carbonate alkalinity; in A′B′C′, the conditions were the opposite: high chloride, hypercapnia, hyperoxia, low carbonate alkalinity.

For example, starting from R of fig. 7.1, the decrease of ambient $[Cl^-]$ (at constant $[Na^+]$) induced a metabolic alkalosis. Then the hypocapnia and the increase of ambient alkalinity which buffers the excreted CO_2

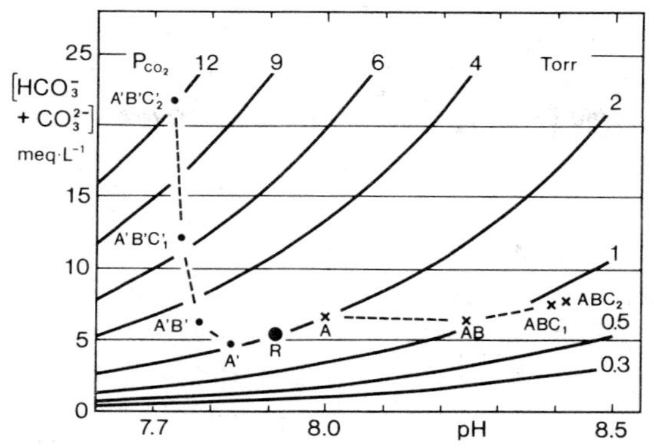

Fig. 7.1 Acid–base balance of the prebranchial hemolymph in the crayfish *Astacus leptodactylus* exposed to various physicochemical conditions of the ambient water at 13 °C (see table 7.2) (Dejours and Armand, 1982).

induced a hemolymph hypocapnic alkalosis (AB); finally the replacement of normoxic water by hypoxic water induced a hyperventilation and a small hypocapnic alkalosis (ABC_1, after a few days; ABC_2 after two more weeks of exposure). Reciprocally it is easy to explain the variations of the hemolymph acid–base balance when the animals are exposed to acidifying environmental changes.

At the end of the experiment, the extreme pH values in the hemolymph were respectively 8.42 (with $P\bar{v}_{CO_2} = 0.86$ Torr and $[CO_3^{2-} + HCO_3^-] = 7.6$ meq · L^{-1}) in the case of the alkalotic series ABC, and 7.73 (with $P\bar{v}_{CO_2}$

TABLE 7.2

Partial composition of the waters to which the groups of crayfish were progressively exposed (R, A, AB, ABC; R, A′, A′B′, A′B′C′). The acid–base balance of the various waters were regulated by pH-CO_2-stats, as described by Dejours and Armand (1982).

		R	A	A′	AB	A′B′	ABC	A′B′C′
$[Cl^-]$	(meq · L^{-1})	1.0	0.25	5.20	0.25	9.70	0.25	9.70
$[HCO_3^- + CO_3^{2-}]$	(meq · L^{-1})	5.0	5.0	5.0	10	0.5	10	0.5
P_{CO_2}	(Torr)	0.80	0.80	0.80	0.30	2.0	0.30	2.0
pH		8.39	8.39	8.38	9.07	7.00	9.07	7.00
P_{O_2}	(Torr)	155	155	155	155	155	40	550
$[Na^+]$	(meq · L^{-1})	5.0	5.0	9.73	10.0	9.73	10.0	9.73

= 12.2 Torr and $[CO_3^{2-} + HCO_3^-]$ = 21.8 meq · L^{-1} in the acidotic series A′B′C′. The acidotic and alkalotic animals were in good shape, had a O_2 consumption slightly above its value during the reference period R, and recovered well when they were returned to the reference water.

This experiment shows that the extracellular acid–base balance may be manipulated by changing the environmental characteristics. Yet it does not mean that it is not regulated, since it may be that it is regulated at different set points according to the ambient milieu. It does not necessarily imply a change of the intracellular acid–base balance which, in the rare cases it has been studied, can remain constant in the face of important environmental changes (p. 37).

Water loss

Terrestrial animals lose water by evaporation unless they live in a completely water-saturated atmosphere. Water evaporation is low if the skin is relatively impermeable to water and the pulmonary ventilation is relatively small; the mechanism and importance of this phenomenon are discussed in Chapter 9. Necessarily, terrestrial animals lose some water by evaporation through the skin, by breathing and through urine and/or feces. The loss of water by respired air interferes with respiration itself in the two examples given below.

(1) Insects lose or gain most of their CO_2, O_2 and H_2O by the spiracular openings. If deprived of water, many insects, mainly those in the larval stage, close their spiracles, reduce water loss but become asphyxic (hypoxic and hypercapnic). Eventually the spiracles will open slightly and will 'flutter', offering tiny apertures through which some O_2 can enter the tracheal system. But since carbon dioxide may have accumulated in the tracheal gas and be mainly buffered in the body fluids, the CO_2 output to the environment may be negligible. In fact there is a small bulk flow of air inwards due to the continuous use of oxygen, which prevents any output of CO_2 and water vapor. This apneic oxygenation cannot last forever, because CO_2 accumulates. From time to time the spiracles open wide and large amounts of CO_2 and H_2O are lost to the environment, while the O_2 stores are replenished. The spiracles then close again. These intermittent openings of the spiracles which allow gas exchange are often called respiratory bursts (see Edney and Nagy, 1976; Bridges and Scheid, 1982;

Schmidt-Nielsen, 1983). The main advantage is that xeric animals save water, as demonstrated in experiments in which the mechanical prevention of spiracular closure results in a very marked water loss. Burkett and Schneiderman (1974) studied the mechanism of the spiracular behavior in the diapausing pupae of the silkworm *Hyalophora cecropia* and of the respective role of hypoxia (mainly ganglionic) and of hypercapnia (mainly peripheral), in the control of the spiracular muscles.

(2) Some mammals – primates, cattle, Camelidae, Equidae – possess sweat glands and their heat loss can be by way of cutaneous water evaporation.

Brooding rock pigeons, and presumably the white-winged doves and the sand grouse, use cutaneous evaporation during egg incubation as a major cooling mechanism (Marder and Gavrieli-Levin, 1986). The cooling of the skin of the brooder aims at creating a thermal gradient between the egg and its surrounding so that the heat produced by the embryo can be eliminated.

But in most mammals (*e.g.* carnivores) and in birds, in general, heat loss results from the regulation of water evaporation in the airways. Within certain limits, ventilation can be increased without an overventilation of the gas exchanging areas, which would result in a hypocapnia and an alkalosis. There are two mechanisms for arriving at this result. (a) Gular flutter, observed in some birds, consists in rapid oscillations of the buccal floor and the throat. (b) Panting, observed in some birds and mammals, and also in some reptiles (see p. 115) consists in a high-frequency, shallow breathing. Some birds, *e.g.*, the cormorant, owl and pelican, respond to heat stress by simultaneous panting and gular flutter (Bartholomew *et al.*, 1968).

In the above cases, the total ventilation and the rate of evaporated water are increased, but the ventilation of the pulmonary parenchyma may be normal; that is, the dead space ventilation may be increased without overventilating the mammalian alveolar lung or the avian parabronchial lung. A normal pulmonary ventilation is preserved, as is shown by the absence of hypocapnia and alkalosis in the arterial blood, at least when the heat stress is not too great. For example, Bouverot *et al.* (1974) observed that awake ducks tolerate exposure for 2 h at 35 °C without hypocapnic alkalosis (see table 7.3). Their ventilation may increase near 7 times over between 20 and 35 °C; but the frequency is increased by 26 while the tidal volume is decreased by 4. But does this

TABLE 7.3

Respiratory values obtained in awake male ducks after 1–2 h equilibration at different ambient temperatures. The head was exposed to a relatively dry atmosphere; the body was at the same temperature as the head, but in a completely saturated atmosphere (from Bouverot *et al.*, 1974).

T air (°C)	20	25	30	35
Number of animals	4	3	4	4
Number of experiments	18	6	13	11
T colonic (°C)	41.4	41.4	41.6	41.7
$\dot{M}_{O_2}$ (mmol · min^{-1})	1.60	1.48	1.71	1.75
f (min^{-1})	10	38	210	260
VT (L BTPS)	0.08	0.066	0.022	0.020
$\dot{V}$ (L BTPS · min^{-1})	0.80	2.51	4.62	5.20
$\dot{V}/\dot{M}_{O_2}$ (L · mmol^{-1})	0.50	1.70	2.70	2.97
$Eair_{O_2}$	0.27	0.080	0.050	0.046
Pa_{CO_2} (Torr)	30.4	29	28.4	27.5
pHa	7.50	7.51	7.52	7.52

hyperventilation increase the energy metabolism, the O_2 consumption of the respiratory muscles and the heat production? If so, the change of breathing pattern might not help to cool the body. Table 7.2 shows that the O_2 consumption does not increase significantly. Perhaps an economical hyperactivity of the respiratory muscles can be achieved because the frequency of panting is close to the resonant frequency of the mechanical respiratory system (Crawford, 1972b; Bartholomew *et al.*, 1968; Schmidt-Nielsen, 1983). However, if the ambient conditions are very unfavorable for heat loss, hyperventilation resulting in hypocapnic alkalosis may occur (Calder and Schmidt-Nielsen, 1968).

CHAPTER 8

Hierarchy of regulations

Summary

The proper oxygenation of the body is in general preserved, at the expense of the CO_2 clearance and the acid–base balance which can be postponed for some time. If oxygenation cannot be maintained and if the recourse to anaerobic metabolism cannot ensure a steady energy production, then the activity of the animals decreases, a phenomenon which is common among invertebrates and in some vertebrates, even mammals (estivation, wintering over, hibernation). The decrease of energy production may be fatal to many animals, but particularly to some fish, birds and mammals.

If the body oxygenation is impaired, some vital organs, the lung, the heart, the brain and probably the adrenal glands tend to be properly oxygenated at the expense of other parts of the body: abdominal viscera, muscles. Several devices, mechanical, chemical, circulatory, tend to maintain the integrity of cerebral functions. Brain sparing is evident in upper vertebrates, but examples are found among fish, molluscs and insects.

In the living organism, thousands of regulations interconnect the various functions. All are necessary for the survival of the individual and the perpetuation of the species. Some functions occur intermittently, or may be postponed or delayed; *e.g.* feeding, reproduction. Others, like respiration, cannot be suspended for more than a few tens of seconds, possibly some minutes or more in poikilotherms. In some rare groups of invertebrates respiration may indeed be arrested, a case of cryptobiosis. Except for this unique situation, energy must absolutely be produced. Energy production may be anaerobic for some time, and some intestinal parasites can live mostly at the expense of their host. All other organisms can turn to anaerobic metabolism for a limited period of time, the duration of which depends on the animal group, on the temperature and on the stage of development. Regarding this last factor, it is well known that embryos of amphibians and of birds, as well as the mammalian fetus, can withstand hypoxia for a longer time than the newborn (Adolph, 1969, 1979). And in turn, survival in hypoxia is longer in the mammalian newborn than in the adult.

But anaerobic metabolism or decrease of metabolism at an early stage of development cannot last forever, and the animals must turn to respiration. Certainly respiration is the paramount function among all those in the body. Many organs may cease to function, and the regulations they ensure may be suspended for some time; yet life is not endangered. For example, disease or experiment may stop the formation of urine for several hours or days, and complete recovery is possible.

Oxygenation, carbon dioxide clearance or acid–base balance?

As is well known, respiration is a complex function. It embraces (1) many subsystems, interposed between the surrounding and the cell; (2) two chemical species, oxygen and carbon dioxide, plus all the factors of acid–base balance in the body fluids, since CO_2 in watery phases is a weak acid. Is there any dominant function (and regulation) in the three components of respiration: O_2 delivery, CO_2 clearance, acid–base balance? The answer to this question, given in Chapters 4 and 5, is straightforward: it is the O_2 delivery which is predominantly ensured. Table 4.2 and fig. 4.2 concern animals in normoxia, that is by definition a condition where the ambient P_{O_2} value is about 150 Torr. One may see that the blood P_{CO_2} value is systematically higher in the air breathers than in the water breathers (compared at the same temperature). It is systematically higher because the air breathers, which for a given O_2 tension disposes of a medium richer in oxygen than water breathers, breathe much less than the water breathers (fig. 5.4). A hypoventilation, whatever model of respiration approximates the reality, leads to an increase of the inspired–expired CO_2 difference and to an increase of the body CO_2 tension. But the CO_2 tension increase does not imply an acidosis because it may be, and actually is, compensated by a proportional increase in bicarbonate. The dual breathers, that is those animals which exchange O_2 and CO_2 in water and in air, either simultaneously or intermittently, illustrate the above contention (table 4.4 and fig. 4.4).

Among the most interesting animals are the amphibians which as

TABLE 8.1

Respiratory characteristics of the trout *Salmo gairdneri* (Randall and Cameron, 1973), the tadpole and adult bullfrog *Rana catesbeiana* (Erasmus *et al.*, 1970/71, at 20 °C; Just *et al.*, 1973, at 23 °C) and the snapping turtle, *Chelydra serpentina* (Howell *et al.*, 1970). The values in parentheses concern a freshwater turtle *Pseudemys scripta elegans* (Jackson, 1971).

	Trout	Bullfrog		Bullfrog		Turtle
		tadpole	adult	tadpole	adult	
Temperature (°C)	20	20	20	23	23	20
$\dot{V} \cdot \dot{M}_{O_2}^{-1}$ (L · mmol^{-1})	5.6					(0.85)
E_{O_2}	0.747					(0.14)
P_{CO_2} (Torr)	2.42	1.95	13.4	4.36	18.9	25.2
pH	7.80	7.83	7.90	7.80	7.70	7.76
$[HCO_3^-]$ (meq · L^{-1})	4.63	4.0	32	8.0	27.5	49.0

tadpoles are water breathers and as adults are air breathers. Table 8.1 gives the data obtained in two experimental series in the bullfrog. It is clear that the blood acid–base balances of the tadpole and the trout are similar. In the adult bullfrog, P_{CO_2} and carbonate concentration are several times higher than in the tadpole, so that the pH values are similar at both stages. Figure 8.1 illustrates this fact for the animals at 20 °C.

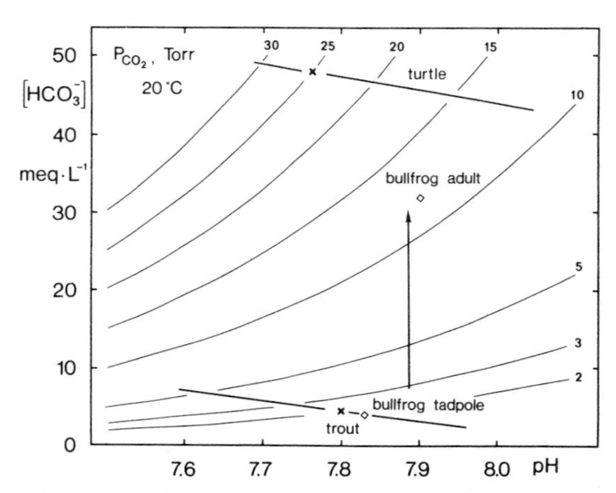

Fig. 8.1. Acid–base balance of blood of trout, tadpole and adult bullfrog and turtle on a $[HCO_3^-]$ *vs* pH diagram. The data are taken from table 8.1. Buffer lines are arbitrary (modified from Erasmus *et al.*, 1970/71).

In favor of the preeminence of bodily oxygenation among the organismal regulations is the observation that hypoxia may enhance the structural development of the gas exchangers: skin, gills, lungs, and increase the blood O_2-carrying capacity, the myoglobin concentration, capillarization, and maybe the enzymatic equipment. However, there is no clear general scheme to be proposed here because the observations vary with species, with duration of the exposure to hypoxia or hyperoxia. For a critical review of this problem, see Bouverot (1985). In case of prolonged hypercapnia, no permanent change, in particular of a structural nature, has been reported.

A second argument for the preeminence of bodily oxygenation is the fact that pure anoxia, as nitrogen breathing, leads to loss of consciousness quickly, the time delay depending on the temperature, the animal species and the body size. A human breathing nitrogen would lose consciousness within 30 to 40 sec, and although he might recover, there is the possibility of permanent nervous damage. On the other hand, marked hypercapnia, for example breathing 30% carbon dioxide in oxygen, will lead, after a period of very unpleasant dyspnea, to loss of consciousness, but the recovery is complete with no enduring disorder. Carbon dioxide narcosis has been used for some years in the treatment of certain psychiatric diseases (Meduna, 1950).

The idea that oxygenation is the primordial aim of respiration, and the CO_2 clearance and the acid–base balance are subordinate, should be accepted without too many difficulties since even a normoxic water, normoxic in terms of P_{O_2} ($\simeq$ 150 Torr), does not contain much oxygen. It is easy to accept that respiration in water breathers is very much oxygen-dependent.

For air breathers, it is probable that the statement of the priority of oxygenation will be questioned. This reticent attitude may derive from what we know of air breathers at sea level which, relative to water breathers, are little sensitive to small changes of oxygen tension and concentration, although an O_2-ventilatory drive may be shown to exist if one carefully observes the ventilatory reactions to variations of the ambient oxygenation (p. 64).

However, air breathers can ventilate much more air than they do at sea level. When they are exposed to high altitude hypoxia, the ventilation increases markedly, leading to a hypocapnia and an alkalosis. The hypoxic hyperventilation is quite obvious in fig. 5.4. Figure 8.2 shows

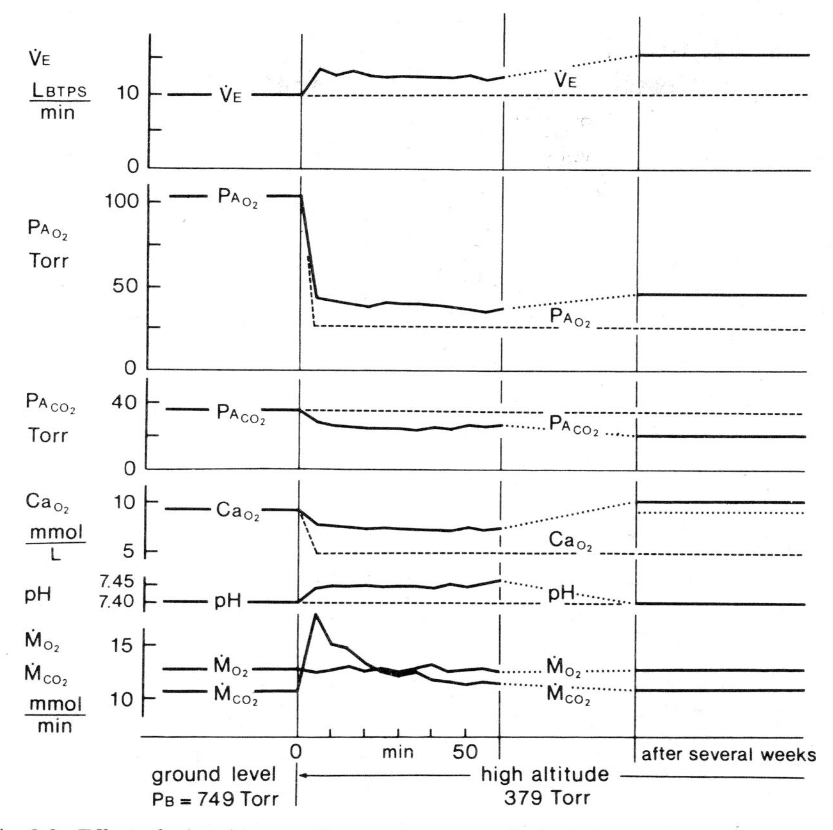

Fig. 8.2. Effect of a low barometric pressure on ventilation, alveolar gas composition, O_2 concentration and pH in arterial blood, O_2 consumption and CO_2 production in humans. The original data at sea level and during the first hour of exposition to high altitude are taken from Rahn and Otis (1947). The values after several weeks of high altitude adaptation are taken from various authors, in particular from Hurtado (1964).

some data obtained in humans exposed to a barometric pressure of 379 Torr (elevation $\simeq$ 5500 m, $P_{I_{O_2}} \simeq$ 70 Torr). One may see that acute hypoxia leads to a hyperventilation with a considerable decrease of the alveolar P_{CO_2} and a hypocapnic alkalosis. The oxygen consumption is not changed, whereas because of the hyperventilation carbon dioxide is washed out, and the CO_2 output and the respiratory quotient increase. After a long delay, the CO_2 output, $\dot{M}_{CO_2}$, returns to its sea level value, that is when the CO_2 stores have reached a new equilibrium. Several weeks later, as fig. 8.2 shows, the hypocapnic alkalosis is compensated. At extreme high altitude in the Himalaya, much lower P_{CO_2} and higher pH have been observed. West reported a value of 7.5 Torr in the alveolar gas of a human breathing air at the summit of Everest. If one considers that the first air breathers probably breathed a very hypoxic air, taking as reference today's sea level P_{O_2} value ($\simeq$ 150 Torr), it is quite possible that they breathed much more than the modern sea level air breathers and had a low P_{CO_2} value. Later on, with the increase of atmospheric P_{O_2}, ventilation could decrease, and P_{CO_2} increase, with no change of pH because of an increase of $[HCO_3^-]$. It is exactly what happens in water breathers which become air breathers, as the amphibians. In terms of evolution, air breathers compared to their aquatic ancestors are in a state of compensated hypercapnic acidosis.

Mechanisms of regulation

Chapters 5 and 6 should be consulted as to the possible mechanisms of regulation which drive ventilation (and circulation). These chapters show that only mammals have been extensively studied. Among the water breathers, the regulation of breathing in various species of crayfish, but mainly in *Astacus leptodactylus*, has been studied (Massabuau *et al.*, 1980, 1984, 1985). The crayfish may be considered as a model for the study of the mechanisms regulating breathing, and what is known in the annelid lugworm and in some fishes fits the crayfish model well.

One important point concerns the acid–base balance. The CO_2 tension and the acid–base balance may most probably be upset by changes of the ambient oxygenation. In some conditions, the change of the acid–base balance may be permanent; but in most animals acid–base balance, as expressed by pH, may return to normal, that is when the bicarbonate concentration is altered in proportion to the change of P_{CO_2} (fig. 8.3). To effect this alteration, the organism disposes of mechanisms – branchial, renal, cutaneous – to regulate the concentration of bicarbonate and of certain other ions as chloride. Anyway there are examples when the blood pH is oxygen-dependent, being permanently lower in hyperoxic depression of breathing, and permanently higher in chronic hypoxia. This situation, together with certain animals whose blood diverges considerably from the range of pH variations *vs* temperature which constitutes Rahn's rule, are not necessarily against the concept of Reeves and Rahn regarding the role of a constant relative alkalinity. What actually is the most important is the intracellular pH on which the dissociation of

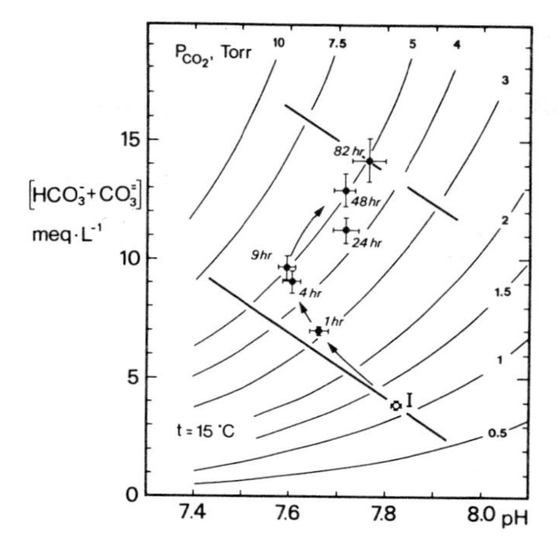

Fig. 8.3. Carbonate *vs* pH changes during emersion of the crab, *Carcinus maenas* at 15 °C. The empty circle and the symbol I indicate the ABB situation during immersion. The filled circles indicate ABB at various times of emersion in hours. The slope of the oblique lines indicates the mean buffer capacity of the hemolymph. Vertical and horizontal bars represent standard errors (from Truchot, 1975).

proteins depends, and a few observations show that intracellular pH may follow Rahn's rule even if the extracellular pH does not (Gaillard and Malan, 1983).

Brain sparing

In upper vertebrates, the mass-specific oxygen consumption of the brain is very high. Because of the size of the brain in adult human, the cerebral O_2 consumption is 20% of the overall O_2 consumption. In the newborn infant, it has been estimated that 70% of the total O_2 consumption is required by the brain (Cross and Stratton, 1974). The cerebral functions appear to be specially protected against perturbations. If the animal is exposed to stressful conditions, mechanisms exist which as far as possible spare the brain, keeping the cerebral functions at their optimum. This concept was clearly expressed in 1911 by Donaldson.

Mechanical protection

The brain is protected against mechanical shocks and vibrations by the skull shield and by its buoyancy in the cerebrospinal fluid, csf, a kind of hydrostatic buffer (Bradbury, 1979). The spinal cord is protected in the same way. Two other vital organs, the lungs and the heart, are mechanically protected by the chest. Certainly there exist some organs which are 'noble', because the cessation of their functions is quickly fatal.

Blood–brain barrier

As all organs, the brain receives blood whose physical and chemical variables are regulated to a certain extent. However, the constituent cells: neurons, oligodendrocytes, astrocytes, are not bathed by blood plasma, but by a small quantity of intercellular or interstitial fluid, isf, which represents about 20% of the brain volume.

The cerebral isf and plasma compositions are very different, and variations of the plasmatic concentration of electrolytes and organic compounds are damped in the cerebral isf, much more damped than they are in the isf of other organs. Apparently the wall of the brain capillaries constitutes a blood–brain barrier which actively secretes the cerebral isf. In the choroid plexuses where the cerebrospinal fluid is formed, similar secretory activity takes place, whence the concept of a blood–csf barrier. The composition of csf and isf is presumably alike and there is no important barrier between these two fluids. The brain having no lymphatic vessels, the isf is presumably drained by csf which itself is drained by the arachnoid villi. There is no blood–brain barrier to oxygen and molecular CO_2, but the mechanism of action of CO_2 on the respiratory centers is much disputed because of the two ionic forms H^+ and HCO_3^- in equilibrium with H_2CO_3 (see Crone and Lassen, 1970; Bradbury, 1979; Kuffler *et al.*, 1984).

In insects, the importance of the blood–brain barrier, actually a barrier between the hemolymph and the brain, is paramount because the hemolymph composition may vary considerably in this animal group (Bradbury, 1979). A 'blood–brain barrier' organization has been reported for the cuttlefish, *Sepia officinalis* (Cserr and Bundgaard, 1984).

Certainly one can consider that the regulation of the cerebral isf composition by the special properties of the blood–brain barrier protects the nervous tissue against wide changes in plasma composition. But it has also been suggested (Crone, 1971) that one function of the blood–brain barrier aims at keeping the neuromediators in the isf, that is to avoid their diffusion into the blood stream.

Brain metabolism

The main substrate for energy production of the brain is glucose. Its transfer from the blood to the cerebral isf is facilitated (Crone, 1965). One might think that hypoglycemia would lead to a compensatory increase of the cerebral blood flow, cbf, so that the glucose delivery to the brain would be maintained constant. Actually, except in very marked and threatening hypoglycemia, cbf is not increased by hypoglycemia; one observes instead an increase of the extraction coefficient of glucose from the blood (Crone, 1965). Finally, the brain may substitute ketone bodies, as β-hydroxybutyrate and acetoacetate, for glucose (Siesjö, 1978).

A striking example of brain sparing is given by the study of the brain development in undernourished weanling rats. At the time of weaning, the brain and body mass of underfed rats were respectively 88% and 50% of control values. Thus the brain in these conditions appeared able to maintain a near-normal development. The incorporation of amino acids, in particular tyrosine and lysine, in the brain tissue was almost normal in spite of a general insufficiency in the rest of the body (Freedman *et al.*, 1980).

Another example of metabolic brain sparing is reported by Childress and Somero (1979) who studied the relationships between some enzymatic activities, the $\dot{M}_{O_2}$ value and the minimum depth of deep-living pelagic marine teleosts. In these animals who inhabit O_2-poor waters (p. 26), the O_2 consumption is less than in epipelagic animals and the activities of four enzymes engaged in respiratory metabolism (lactate dehydrogenase, pyruvate kinase, malate dehydrogenase and isocitrate dehydrogenase) are decreased in skeletal muscles. However, the activities of these enzymes remain constant in the brain and in the heart. These differences appear to reflect the 'aerobic poise' of these organs.

Cerebral blood flow and oxygen consumption

Taking as reference the conditions prevailing at sea level, changes of the O_2 and CO_2 status in blood perfusing the brain induce very important variations in the cbf. In mammals and humans, hyperoxia decreases cbf; hypoxia increases it. On the other hand, hypercapnia increases it, whereas hypocapnia decreases it. In both cases the variations are quite marked. They may be viewed as protective regulatory reactions since they minimize locally the variations of the O_2 and CO_2 concentrations.

The situation, however, is not very clear, because any natural change in cbf must lead to simultaneous changes of oxygen and carbon dioxide in the brain. For example, cbf rises at high altitude; but high altitude entails also a hyperventilation and a hypocapnia. Since by itself, hypocapnia lowers cbf, it is obvious that the change of cbf at high altitude is a complex phenomenon resulting from an increased effect of hypoxia and a decreased effect of hypocapnia. Indeed, an experimental isocapnic fall of P_{O_2} increases cbf more than a fall of P_{O_2} accompanied by a hyperventilation-induced hypocapnia (Siesjö, 1978). The interpretation is further complicated by the Bohr–Haldane effect on blood, according to which one cannot change the conditions of either O_2 or CO_2 without changing the other. Furthermore, as noted above, a durable change of P_{CO_2} in blood is generally metabolically compensated. What happens to cbf in a subject chronically exposed to high altitude? Presumably his cbf increases concurrently with the progressive compensation of the hypocapnic alkalosis. On the other hand, prolonged exposure to high altitude hypoxia, at least in most mammals, entails an increase of [Hb] which is commonly considered as adaptive, and it is known that a Ca_{O_2} increase (at constant Pa_{CO_2}), which would be the result of a higher [Hb], induces a cbf decrease.

In spite of an abundant literature on cerebral metabolism, respiration and circulation, the data are quite evidently far from complete (Busija and Heistad, 1984; Siesjö, 1978). In particular the observations in lower vertebrates are very scarce. In any case it remains that in higher vertebrates, notably in normal humans, the oxygen consumption of the brain is remarkably constant down to very low P_{O_2} (Siesjö, 1978).

Against hypercapnia and hypocapnia, the cells of the brain, like any cells, are relatively protected by their high buffering capacity. However, probably, the brain tissue has special properties against acid–base disturbances. When the arterial P_{CO_2} of dogs is varied between 15 and 85 Torr, the cerebral cortex undergoes compensatory changes which return intracellular pH to normal within 3 h, when arterial and csf pH values are still very far from their control values (Arieff *et al.*, 1976).

In the *fetus*, the blood which reaches the head contains more oxygen than the blood which goes to the lower part of the body (Barcroft, 1938). This is possible because the relatively well-oxygenated blood returning from the placenta via the umbilical veins reaches the right auricle and passes directly into the left auricle through the foramen ovale, and from there perfuses the upper part of the body. Most of the less well oxygenated blood of the superior vena cava passes to the right ventricle, the pulmonary artery and reaches, via the ductus arteriosus, the aorta beyond the orifice of the cephalic arteries (Girard *et al.*, 1983). In ewes rendered hypoxic by breathing hypoxic mixtures, one observes in the fetus a redistribution of the blood flow through its organs (Peeters *et al.*, 1979). In the hypoxic fetus, the blood flow to the brain, mainly the cerebellum and brain stem, to the myocardium and the adrenals increases considerably so that the amount of O_2 carried to these tissues remains constant. Concomitantly the amount of O_2 delivered to the viscera and the carcass decreases. Analogous results were reported by Behrman *et al.* (1970) in the macaque. This redistribution of blood in favor of the brain is presumably a live-saving reaction to the asphyxic stress which occurs at birth.

Brain protection is also observed in prolonged *dives* which some pinnipeds, as the Weddell seal, undertake from time to time. If in natural dives lasting less than 20 minutes this seal is able to ensure the production of energy aerobically by the use of the important O_2 stores it carries along, the situation is completely different for dives which may exceed one hour. In long dives, as demonstrated first in 1940 by Scholander (see Scholander, 1964), the blood flow is redistributed. The brain circulation is privileged while most of the other parts of the body (1) turn to lactic acid fermentation, and (2) decrease their metabolic production (Elsner and Gooden, 1983). In simulated dives of the Weddell seal lasting 8–12 min (Zapol *et al.*, 1979), the cardiac output decreased from 39.8 to 5.6 $L \cdot min^{-1}$, the cardiac frequency fell from 52 to 15 $beats \cdot min^{-1}$ and the blood flow to the splanchnic viscera and to the carcass was reduced by more than 90%. The cerebral blood flow remained unchanged; the myocardial blood flow decreased in proportion to the fall of the cardiac output. The adrenal circulation decreased moderately.

In *birds*, a similar phenomenon of cerebral protection against asphyxia has been reported. In ducks and the bar-headed goose breathing a 5% O_2–95% N_2 mixture, a change of the blood flow pattern was observed; the blood flow to the brain, adrenal glands, heart and eyes increased, whereas the blood flow to splanchnic viscera decreased (Faraci *et al.*, 1985).

Pressure of the blood perfusing the brain

In order to ensure a proper cerebral blood flow, a satisfactory blood pressure in the carotid and vertebral arteries is required. This pressure is relatively independent of (1) animal size, and (2) posture. (1) In human, ox and giraffe, the mean arterial blood pressure at the entry of the skull varies between 75 and 100 Torr, a range of pressure which is also observed in smaller mammals. (2) The blood pressure at the level of the heart increases by 15 Torr about where a supine human stands up (Bjurstedt *et al.*, 1962). In the standing giraffe, the mean blood pressure is 95 Torr at the brain level and 215 Torr at the heart level; the difference of 120 Torr approximates the pressure of a column of blood of 155 cm, that is, close to the length of the neck (Patterson *et al.*, 1965). Consequently the blood pressure in the carotid and vertebral arteries near the skull is relatively independent of the body posture. However, it is difficult to evaluate the role of this independence in the absence of values for jugular venous blood pressure (Badeer, 1986). The circulatory reaction to orthostatism is nonetheless certainly important, since in its absence, people are inclined to faint, a phenomenon which in this case is attributed to a fall of cerebral blood flow.

Temperature

In some mammals and birds the brain temperature is noticeably different from the esophageal and carotid artery blood temperature. In particular, it may be higher in mammals at rest in a temperate environment; this is not surprising since the mass-specific cerebral metabolic rate is one of the highest of the body. But cerebral temperature may be lower than blood temperature when animals are exposed to heat stress or perform prolonged heavy exercise in a hot environment. Some examples follow.

In *mammals*, there are some striking examples showing that the brain temperature may be much lower than the core temperature (Baker, 1982). For example, Taylor and Lyman (1972) measured carotid artery and brain temperatures in the gazelle by implanted thermocouples. In running gazelles, the brain temperature did not increase as rapidly as the temperature of the blood in the carotid artery. For example, after a 7-min run at $40\,km \cdot h^{-1}$, carotid temperature was close to 44 °C whereas the brain temperature was only 40.5 °C. It seems inescapable to see this behavior as being favorable for cerebral function. Apparently, the blood supplying the brain is cooled via a countercurrent heat exchange in the carotid rete with cool blood draining the nasal mucosa.

In *humans*, the brain may be cooled by the venous blood returning from the facial skin via the ophthalmic vein to the cavernous sinus, where the arterial blood perfusing the brain may lose some heat. It has been shown, indeed, that face fanning decreases the temperature of the forehead skin, the angularis oculi vein blood and the tympanum (McCaffrey *et al.*, 1975; Cabanac and Caputa, 1979).

Birds, even at rest, maintain cerebral temperatures 1 °C below body core temperature (Baker, 1982). The carotid arterial blood which perfuses the brain is cooled in the *rete ophthalmicum* which exists in all birds studied (Midtgård, 1983). In the *rete* the carotid artery divides into small arteries and intertwines with veins which return a relatively cooled blood from the bill, the eye and the evaporative surface of the upper airways. The flows of the arterial and venous vessels run countercurrently, so that the carotid blood streaming along the retia is cooled by the venous blood before reaching the brain. In the pigeon, which has been extensively studied, Pinshaw *et al.* (1985) suggested that besides their cooling function, the retia might be the seat of gas exchange such that the O_2 supply to the brain is enhanced.

Some *reptiles* pant when exposed to heat. In particular, in the chuckawalla *Sauromalus obesus* exposed to an ambient temperature of 45 °C, panting maintains deep body temperature at 44.1 °C, but the brain temperature is kept at 42.3 °C. The blood of the carotid arteries is cooled by the evaporation at the surface of the pharynx (Crawford, 1972a; Schmidt-Nielsen, 1983, p. 285).

CHAPTER 9

From respiration to other functions in aquatic and terrestrial animals

Summary

The striking contrasts in the respiratory characteristics of water- and air breathers are related to the relative superabundance of oxygen in air. But almost all the physical properties of water and air differ, inviting an extensive comparison of the biological traits of aquatic and terrestrial animals.

All terrestrial animals had to develop mechanisms for conservation of water and concentration of salts and nitrogen wastes. Generally wholly aquatic animals are ammoniotelic, whereas active terrestrial animals are uricotelic or ureotelic. In relation with the constraint of gravity, the mass percentage of supporting tissues, skeleton, cuticle, eggshell, increases with size in terrestrial animals. Terrestrial animals have some particular circulatory features because they lack the buoyancy available to the aquatic animals. Although there are a few examples of organ homeothermy in fish, only upper vertebrates are completely homeothermic because the dissipation of heat in air is much lower than in water. Whales and seals, which descend from terrestrial ancestors, are apparent exceptions since their homeothermic core is reduced.

Whether the aquatic animals are invertebrates or vertebrates, the difficulties of the invasion of land are common to all. Many biochemical, anatomical and physiological traits are convergent and are related to the characteristics of the environment. Amphibious animals, such as the amphibians during their metamorphosis, may offer models to understand evolution from aquatic to terrestrial life.

In the foregoing chapters the consequences of the properties of water and air for the status of oxygen and carbon dioxide in the body fluids have been studied. We have seen that the air breathers breathe less than water breathers because they have at their disposal a O_2-rich medium, and that, consequently, their body fluid P_{CO_2} values are relatively high; however, the pH values are about the same (for the same temperature) in the two groups of animals. If we take the air breathers as the heirs of the water breathers, their acid–base balance corresponds to a compensated hypercapnic acidosis. These considerations are reinforced by the study of amphibians which may show during their life time, by metamorphosis from an aquatic tadpole to an air-breathing adult, what our aquatic ancestors experienced some 400 megayears ago.

TABLE 9.1
Main differences between water and air and their physiological consequences.

	Water	Air	Air/water	Consequences
O_2 and CO_2 diffusivities	+	++++	≃ 8000	O_2 & CO_2 tensions acid–base balance
O_2 capacitance	+	++	≃ 30	
CO_2 capacitance	++	++	≃ 1	
NH_3 capacitance	++++	++	≃ 1/700	N end products
Viscosity	++	+	≃ 1/60	work of breathing circulation
Density	+++	+	≃ 1/800	skeleton locomotion
Kinematic viscosity	+	++	≃ 13	buoyancy, gravity
Water availability	+++	very variable		water turnover osmoregulation ionoregulation
Ionic environment	very variable			
Sound velocity	++	+	≃ 1/4	audition
Sound absorption	++	+	≃ 1/4	
Light refractive index	1.33	1	≃ 0.75	vision
Light absorption	++	+	≃ 1/12	
Dielectric constant	+++	+	≃ 80	electroreception
Solubility of molecules	variable			distance chemoreception
Volatility of molecules		+		
Diffusivity of molecules	+	++++		
Heat capacity	++++	+	≃ 1/3500	heat dissipation body temperature
Heat conductivity	++	+	≃ 1/24	
Heat of evaporation	≃ 2450 kJ · L^{-1}			

Table 9.1 shows the main physicochemical differences between water and air. The heading 'water' includes all waters: fresh, brackish, sea and briny waters. The differences in ionic composition, titration alkalinity and osmolarity between these waters have important consequences as to the relevant functions. The soil environment is special (p. 28). In many respects it is intermediary between water and air, having respiratory consequences similar to water when it is very wet or inundated and similar to air when it is dry. Since little is known about the physiology of dwellers in different soils, for simplicity and also for lack of pertinent data, no soil column appears in table 9.1.

Most functions, if not all, can be affected by the different characteristics of water and air (Bliss, 1979). Is there any correlation between the various physiological changes imposed by the properties of the milieus? Or should the contrasts between terrestrial and aquatic animals be studied one after the other, juxtaposed as in a catalogue? We will see that there may be a logical sequence of presentation of these contrasts, which is of interest if it approximates the time course of biological evolution. It is arbitrary, albeit practical, to study one function after another. What must always be kept in mind is that all functions of the body must be satisfied together in order to ensure the survival of individuals and the perpetuation of the species.

Nitrogenous end products

For animals excreting ammonia to emerge from water and colonize open lands, the nature of the end products of nitrogen-containing organic molecules had to change. Ammonia is very toxic, but since it is extremely soluble in water, the NH_3-NH_4^+ system is kept at low concentration in the body fluids of the water breathers. In air, NH_3 is comparatively not very 'soluble'; its capacitance is equal to $1/R\mathrm{T}$, as for any gas species as O_2 and CO_2 (p. 22). Because it is toxic, NH_3 cannot be accumulated in the body fluids of air breathers at a concentration at which it could be eliminated as such. Consequently the terrestrial animals excrete less toxic products, mainly urea and uric acid, in order to get rid of organic nitrogen.

Water breathers

Most water breathers are ammoniotelic. This is possible because water is a sink for the extremely soluble ammonia.

According to the Brønsted–Lowry definition of acids and bases

$$NH_4^+ \rightleftharpoons NH_3 + H^+ \tag{9.1}$$

with a dissociation constant, K'_{am} defined as:

$$\frac{[NH_3]\,[H^+]}{[NH_4^+]} = K'_{am} \tag{9.2}$$

Hence,

$$pH = pK'_{am} + \log \frac{[NH_3]}{[NH_4^+]} \tag{9.3}$$

For example, in distilled water at 15 °C, $K'_{am} = 3.31 \cdot 10^{-10}$ and $pK'_{am} = 9.48$. The pK'_{am} value as well as the NH_3 solubility varies with temperature and salinity (ionic strength) (Cameron and Heisler, 1983). The ambient water pH is generally much lower than the pK'_{am}, so that $[NH_3] \ll [NH_4^+]$. For example, if the pH value of a freshwater at 15 °C is 7.48, $[NH_3]$ is 100 times less than $[NH_4^+]$.

Because NH_3 is very soluble, its partial pressure in ambient water is very low, 10 μTorr (table 9.2, column A, line a). The amount dissolved is 0.475 $\mu mol \cdot L^{-1}$ (column A, line c). In venous plasma which reaches the gills the value of P_{NH_3} and C_{am}tot are 87 μTorr and 393 $\mu mol \cdot L^{-1}$ (column D, lines a and i). During the branchial exchanges some ammonia moves from the blood to the water, and arterial P_{NH_3} is as a result 40 μTorr with a total C_{am}, C_{am}tot, of 199 $\mu mol \cdot L^{-1}$ (column C, lines a and i). This simple example does not imply that ammonia is actually transferred from the body fluids to the ambient water by a simple diffusion of unionized NH_3. The mechanism of NH_3 transfer is much disputed, since it may involve NH_3, NH_4^+ and eventually the formation of ammonia from blood glutamine. In the cells, whose pH value is much lower than that of plasma, most of the ammonia is protonated as NH_4^+. This ion may be accompanied by HCO_3^- or there may be an exchange NH_4^+/Na^+. But the

TABLE 9.2

Status of the ammonia–ammonium pair in inspired and expired water, in arterial and venous blood of trout at 15 °C. The values of columns A, C and D are taken from Cameron and Heisler (1983). Column B is theoretical because the exact composition of the expired water is unknown; however, this theoretical example shows that P_{NH_3} is very low in this water; the fall of P_{NH_3} between inspired and expired water is due to the fact that the expired water being more acid than the inspired water (because of the CO_2 output), its NH_4^+ concentration is higher, whereas its dissolved NH_3 concentration and the corresponding P_{NH_3} are lower. The hypothetical example of column E shows the enormous increase of P_{NH_3} that would be necessary for the trout to excrete NH_3 in air instead of water with the same ventilatory flow rate (see text).

	A inspired water	B expired water	C arterial plasma	D venous plasma	E hypothet. expired gas
a P_{NH_3} (Torr)	$10 \cdot 10^{-6}$	$1.7 \cdot 10^{-6}$	$40 \cdot 10^{-6}$	$87 \cdot 10^{-6}$	0.39
b β_{NH_3} ($\mu mol \cdot L^{-1} \cdot Torr^{-1}$)	47 530	47 530	49 450	49 450	55.6
c C_{NH_3} ($\mu mol \cdot L^{-1}$)	0.475	0.081	1.97	4.28	
d pK'_{am}	9.48	9.48	9.64	9.64	
e pH	7.00	6.17	7.64	7.70	
f $pK'_{am} - pH$	2.48	3.31	2.00	1.94	
g $10^{(pK'_{am} - pH)}$	302	2040	100	87	
h C_{NH_4} ($\mu mol \cdot L^{-1}$)	143	165	197	389	
i $C_{am}tot$ ($\mu mol \cdot L^{-1}$)	143.5	165	199	393	21.5

solubility, and consequently the Krogh constant of diffusion of NH_3, is extremely high; it may be that most of the NH_3-NH_4^+ pair is excreted as NH_3. In all milieus, however, eqs. (9.2) and (9.3) hold true (see Cameron and Heisler, 1983; Wright and Wood, 1985).

Ammoniotelism is widespread among water breathers, but there are exceptions. The elasmobranchs, rays and sharks, synthesize mainly urea, to a lesser extent trimethylamine oxide (TMAO) and some ammonia, and excrete all three substances (see, for example, Goldstein and Forster, 1971). The plasma concentrations of urea and TMAO may reach respectively 400 and 200 $mmol \cdot L^{-1}$. The osmotic role of urea and TMAO in the marine elasmobranchs whose plasma osmotic pressure is close to that of sea water is confirmed by the observation that in a freshwater elasmobranch, the stingray *Potamotrygon*, the plasma osmolarity is typical of a teleost fish ($\simeq$ 300 mmol $\cdot$ L^{-1}) with no TMAO and very little urea in the plasma (Thorson *et al.*, 1967).

The general rule of ammoniotelism in water breathers does not imply that their exclusive end product of the nitrogen metabolism is NH_3-NH_4^+. It means only that ammonia is excreted predominantly; some urea, amino acids and purine derivatives may also be. For details of the excretion of nitrogenous end products in marine and freshwater water breathers see, for example, Campbell and Golstein (1972) and Schoffeniels (1984).

Air breathers

Ammonia is several hundred times less 'soluble' in air than in water, *i.e.* the air's NH_3 capacitance is much lower. If air breathers were ammoniotelic, what would the situation be? The animal could have a very high P_{NH_3}; table 9.2, column E, shows that if trout could breathe air, the expired P_{NH_3} would be 0.39 Torr. This would ensure an increase of 21.5 µmol $\cdot$ L^{-1} of the total NH_3-NH_4^+ concentration between expired and inspired gas (column E, line i) which is the increase observed in comparing expired to inspired water in the trout (line i, columns A and B). Since ammonia is very toxic, this is unimaginable (Campbell, pp. 284–288, in Prosser, 1973). On the other hand, the animal could ventilate enormously, a minute volume of air large enough to keep P_{NH_3} low and non-toxic. The example of table 9.1 would mean that several hundred liters of air would have to be ventilated instead of 1 L of water. This solution is precluded by the cost of breathing and the probability of desiccation by evaporation avoidable only by the animal's occupying a water-saturated niche. Even a compromise between the two hypothetical solutions would not circumvent the impossibilities.

Consequently, to colonize the land, the eventual air breathers must have had to change their main nitrogenous end products. First of all, some NH_3, a minute amount, can actually be eliminated by air and more NH_3-NH_4^+ may be excreted with the urine and feces. Some air breathers are in fact ammoniotelic, *e.g.*, the crustaceans, isopods and some amphipods. These animals are small, have a low energy metabolism and a relatively large surface area in contact with the ambient air. Apparently they produce NH_3 directly on their cuticle which contains a glutaminase (Wieser and Schweizer, 1972).

Some large crustaceans are really terrestrial, *e.g.* the land crab *Car-*

disoma guanhumi (Bliss and Mantel, 1968). They may form important amounts of urea, uric acid and some amino acids; uric acid may not be excreted, but deposited in the abdominal hemocoel and the mid gut gland (Gifford, 1968; Horne, 1968; Craybrook, 1983), but most of the terrestrial crustaceans could not develop other routes of nitrogen excretion efficiently (Regnault, 1987). This may be one reason, together with the lack of a proper external gas exchanger, why they were unable to radiate successfully on land, as did the insects and the arachnids.

Some terrestrial snails produce some NH_3-NH_4^+ which may play a role in the deposition of the calcium carbonate of the shell. The alligator excretes a sizable amount of organic nitrogen in the urine as ammonium bicarbonate; the rest of the excreted organic nitrogen is uric acid (Coulson and Hernandez, 1983). But usually terrestrial invertebrates and vertebrates eliminate the bulk of the organic nitrogen as urea, a water-soluble and little toxic product, or as uric acid or other purine derivatives: allantoin, allantoic acid, guanine, *etc.*, which are deposited in the body or excreted. Each animal group has its own characteristic protein catabolism and nitrogenous end products (Campbell and Goldstein, 1972).

Amphibious breathers

In some aquatic larva of *insects* such as *Dytiscus* the nitrogenous end product is ammonia. However, the pupae and imagos turn to uricotelism. Uric acid may be precipitated and eventually excreted (Maddrell, 1971; Wigglesworth, 1972).

In *amphibians*, the transition from aquatic ammoniotelism to terrestrial uricotelism, ureotelism, or a combination of both, is well-documented (see Schmidt-Nielsen, 1983). The biochemical differentiation underlying the transition from ammoniotelism to ureotelism has been studied in *Rana catesbeiana* (Cohen, 1970). The newt, *Notophthalmus viridescens*, generally aquatic, may leave its pond in the spring and summer and return to water in the fall (Walters and Greenwald, 1977). In the newt in air, the blood concentration of total NH_3 does not increase but the urea concentration rises by a factor of 2.5. At this time, the activity of liver arginase, a key enzyme of the Krebs–Henseleit ornithine–urea cycle, increases thirty times. When the newt returns to water, the urea returns quickly to its basal aquatic concentration. Degani *et al.* (1984) studied the green frog in freshwater and after three months of living in dry soil; they observed that the urea concentration in plasma was 20 mmol · L^{-1} in water and reached 900 mmol · L^{-1} in the dry conditions. Two xeric treefrogs are able to produce some uric acid together with urea (see p. 125) (Shoemaker *et al.*, 1972; Campbell *et al.*, 1984).

In conclusion:

(1) nitrogenous end products may be the NH_3-NH_4 pair, urea, uric acid and some other purine derivatives;

(2) generally, one nitrogenous end product is dominant, taking into account all possible pathways of excretion – gills, lungs, urine, feces and even end-product storage. When one says, *e.g.*, that the teleosts are ammoniotelic, it means that the dominant nitrogenous end product (in terms of nitrogen content), but not necessarily the only end product, is the ammonia system;

(3) most water breathers are ammoniotelic;

(4) aquatic animals which are not ammoniotelic but are mainly ureotelic, like sharks and rays, accumulate nitrogen products as urea and TMAO so that the osmotic pressure of their body fluids will be approximately that of the environment;

(5) in animal groups in which some species are aquatic and others are terrestrial, the former are ammoniotelic, the latter are ureotelic or purinotelic;

(6) in groups which start as water breathers and then turn to air breathing, particularly the amphibians, the animals are ammoniotelic during the aquatic stage, and generally are ureotelic during the terrestrial stage. If they return to water, in winter for example, they switch back to ammoniotelism;

(7) ammonia is toxic. However, (a) the toxic concentration varies widely among animal groups, and (b) the mechanism of the toxic action is unclear (see Campbell in Prosser, 1973; Cooper and Plum, 1987).

Water conservation, and excretion of salts and nitrogenous end products

In general, water loss is proportional to metabolism and to the availability of water (Schmidt-Nielsen, 1969; Nagy and Peterson, 1987). An animal able to colonize land, particularly arid lands, must save water and concentrate salts and nitrogen end products to be excreted. Examples will be taken mainly in the class of Amphibians because they live in all possible environments, except sea water, although some can inhabit brackish waters (Gordon *et al.*, 1961; Schmidt-Nielsen and Lee, 1962; Tercalfs and Schoffeniels, 1962).

Aquatic tadpoles do not have to conserve water since they are hyperosmotic to the milieu. If they become terrestrial with metamorphosis, they may avoid desiccation by dwelling in a damp niche. If water is not readily available, they may resist desiccation for some time because (1) they generally can reabsorb the hypotonic urine of their bladder which, in some species, has a large capacity; (2) they may concentrate their body fluids to a certain extent (Bentley, 1966). But amphibians must return to water from time to time to absorb water and to eliminate urea and salts.

The evaporative water loss is generally thought to be enormous in amphibians (Bentley, 1966). However, in some arboreal amphibians living in a dry environment it is relatively low (Wygoda, 1984). The best examples of amphibian water economy are found in two genera of treefrogs, *Chiromantis* (Loveridge, 1970) and *Phyllomedusa* (Shoemaker *et al.*, 1972). These remarkable amphibians have two characteristics which permit them to live in desert areas: (1) their cutaneous water loss is very small and comparable to that of xeric reptiles; (2) they are mainly uricotelic, and since urate precipitates, they need almost no water for its excretion; furthermore urate carries with it a cation which then is disposed of without water. However, when the waterproof frogs *Phyllomedusa* and *Chiromantis* were exposed to extreme heat, a sudden increase of evaporative water loss was observed, due to a secretory activity of the skin, analogous to sweating (Shoemaker *et al.*, 1987).

Urate excretion is also very common in reptiles and in birds, and many eliminate salts together with urates in the feces. However, those species without access to freshwater, *i.e.* marine and desert species, are provided with a variety of salt glands (see Schmidt-Nielsen, 1983). Kidneys of amphibians, reptiles and birds cannot make a urine more concentrated than the plasma, but water can be reabsorbed in the bladder, or in the distal part of the intestinal tract, or both (Shoemaker and Nagy, 1977). Only the mammalian kidney makes a hypertonic urine, up to a concentration of 9 osmol $\cdot$ L^{-1} in a desert rodent, the Australian hopping mouse (MacMillen and Lee, 1969). Schmidt-Nielsen (1964) has shown that some desert rodents are virtually independent of external water.

To sum up, three factors ensure water conservation: (1) decrease of urinary and fecal water loss; (2) a reduced permeability of the integument to water, as observed in a few amphibians (treefrog), wholly terrestrial reptiles, birds, mammals and insects; no integument is in fact completely impermeable to water; (3) the relatively low values of the pulmonary ventilation by which much water might otherwise be lost. We will return later (p. 136) to the importance of the role of reduced ventilation in the occurrence of homeothermy.

The water conservation phenomenon is general among terrestrial animals. For example, most of the evaporative water loss of terrestrial and aquatic brachyuran crabs occurs through the shell (Herreid, 1969a). However, crabs from a terrestrial habitat lose the least water while aquatic species lose the most. Nevertheless, even in the most terrestrial crabs, the water loss is several times greater than that of arachnids, insects and terrestrial vertebrates (Herreid, 1969a), and Powers and Bliss (1983) suggest that an 'inefficient capability for water retention' is presumably the most important factor which limited the colonization of land by decapod crustaceans. We have indicated previously that the relative inefficiency of the external exchanger for oxygen transfer and the incomplete shift from ammoniotelism to ureotelism or uricotelism in terrestrial crustaceans (p. 123) may be also factors which explain why the colonization of land by these animals is restricted to a few genera.

Respiration and acid–base balance

If the problems of nitrogenous end products and water economy are solved, terrestrial animals are in a position to take advantage of the abundance of oxygen in air. We have seen (p. 36) that air breathers breathe much less than the water breathers and that, as a consequence, they are hypercapnic; but since an increase of the body fluid bicarbonate concentration is nearly proportional to the increase of P_{CO_2}, the actual pH values in body fluids of aquatic and air breathers are similar (at a given temperature). We will not return to these phenomena because they have been largely studied beforehand, but a few more words must be said about ionoregulation.

We have seen that in the amphibians the transition from water breathing to air breathing is accompanied by an increase of $[HCO_3^-]$. This increase of [HCO_3^-] should not raise an important problem for amphibians because their cationic and anionic concentrations may vary widely with the environmental conditions to ensure an ionic balance (Shoemaker and Nagy, 1977; Duellman and Trueb, 1986, pp. 207 and 208). But in birds and in mammals, ventilation changes greatly, and consequently so does P_{CO_2}, with the changes of $P_{I_{O_2}}$ due to the barometric variations with altitude. When in humans, for example, P_{CO_2} falls from 40 Torr at sea level to 20 Torr at high altitude, the bicarbonate concentration should decrease by half, *e.g.* from 26 to 13 meq · L^{-1}, and does, if, after acclimatization, the pH is to be the same at the two elevations. According to the observations of Dill *et al.* (1937) the fall of [HCO_3^-] is compensated mainly by an increase of [Cl^-] and to a less extent by a decrease of [Na^+]. The overall concentration of osmolytes remains the same in the two conditions.

Table 4.2 (p. 38) and fig. 4.2 (p. 41) show that the hypercapnia of air-breathing arthropods is not very important, although the observed P_{CO_2} is always higher than the values in water-breathing animals. For example, according to table 4.2, at 20 °C the blood P_{CO_2} value of the terrestrial fiddler crab is 9.7 Torr whereas it reaches 22.7 Torr in the red-eared turtle at the same temperature (p. 39). Relatively modest hypercapnia is also found in the crab *Gecarcinus lateralis*, the coconut crab and the tarantula (table 4.2). Also table 4.5 (p. 48) shows that the emersion of the crab *Carcinus maenas* and of the crayfish *Austropotamobius pallipes* leads to a moderate hypercapnia; even if the emersion is prolonged several days the hypercapnia is never very marked (Truchot, 1975). The reason for the moderate hypercapnia of air-breathing arthropods is not known. But it must be pointed out that the hemolymph P_{O_2} value is rather low – the inspired air-to-blood P_{O_2} difference exceeds 100 Torr – and is lower during air breathing than during water breathing (Truchot, 1987a). It is possible that this hemolymph hypoxia leads to a relative hyperventilation limiting the hypercapnia.

Oxygen-carrying properties of blood

The O_2-carrying capacities of blood or hemolymph differ broadly between animal groups. In general, the concentration of pigment is higher in air breathers than in water breathers. But the blood of some water breathers such as tuna and the lugworm *Arenicola* contains a relatively high concentration of pigment. On the other hand (p. 58), the relation between the specific cardiac output and the arterial blood O_2 concentration is the same in both groups of animals.

What about their blood O_2 affinity? This is expressed by the location of the curve relating the blood O_2 concentration and the O_2 pressure, the so-called oxygen dissociation curve (ODC). A shift to the right of this curve means a decrease of the blood O_2 affinity (which entails an increase of P_{50}, the P_{O_2} value for which 50% of Hb is oxygenated). *Mutatis mutandis* for a shift to the left. Is there any systematic change of the position of the ODC with the O_2 availability in the milieu, in particular between water- and air breathers?

A teleological reasoning suggests that blood of air breathers could have a lower O_2 affinity than the blood of water breathers. Indeed air breathers live in an O_2-rich milieu, where the O_2 diffusivity is high and the unstirred layer of milieu is thin, as compared to water; these factors facilitate the O_2 loading of the blood at the cutaneous or pulmonary surfaces. For animals living in an O_2-poor milieu, it is advantageous to possess a blood with a high O_2 affinity (low P_{50}), because O_2 loading at the gas exchange surfaces is favored. A steep oxygen dissociation curve (high arterial–venous O_2 capacitance) is also favorable. What is the reality in face of these teleological considerations?

Krogh and Leitch (1919) compared two groups of fish: one group of carp, eel and pike which are often exposed to low O_2 pressure and a second group of trout, cod and plaice which live in well-oxygenated water. They reported that the first group had a high O_2 affinity, whereas the second group had a low oxygen affinity.

Is this clear conclusion of Krogh and Leitch confirmed for other aquatic animals and does it hold true for air breathers with their easy access to oxygen as compared to water breathers with their scanty supply of oxygen? Some reports (Johansen and Lenfant, 1972; Wood and Lenfant, 1979; Powers *et al.*, 1979) confirm the view of Krogh and Leitch. Also Ar *et al.* (1977) reported that fossorial animals which live in an O_2-poor atmosphere (p. 28) have a high O_2 affinity. But a review of the literature does not seem to permit any firm conclusion at the present time (Wood and Lenfant, 1979, 1987; Wood, 1987). Certainly one reason for this lack of unanimity is the difficulty of comparing the thousands of data available. Indeed, many factors may shift the oxygen dissociation curves: temperature, ionic concentration, duration of acclimatization, stressful conditions. In some animals, additionally, as in the lugworm *Arenicola* (Toulmond, 1975), the oxyhemoglobin is more an oxygen store than an oxygen carrier playing a permanent role in the loading and unloading of O_2 between the respiratory surfaces and the tissues.

The only general law which seems firmly established, for all vertebrates at least, is that the blood oxygen affinity is lower in animals after their birth than in prenatal stages; this change of O_2 affinity may be due to structural modifications of the hemoglobin molecules, to variations in the concentration of intraerythrocyte modulators (ATP; 2,3-DPG, *etc.*). Of special interest is the decrease of O_2 affinity which accompanies the metamorphosis of a water-breathing tadpole to an air breathing amphibian adult (Wood, 1971; Broyles, 1981).

Are aquatic insects water- or air breathers?

It has been claimed repeatedly that all animals, even the terrestrial ones, are actually water breathers because the surfaces of the gas exchangers, such as the alveolar surface of the lungs, are wet. First, it is not true that all gas exchange surfaces are wet; in small aerial arthropods, O_2 and CO_2 exchanges occur through an essentially dry cuticle. But it is correct that the inside of the lungs, as well as the endings of the tracheae in insects, are wet. Nevertheless, these animals are air breathers because they take up O_2 from and eliminate CO_2 into a gas phase.

Carbon dioxide- and oxygen exchangers in air breathers are of three types: (1) A diffusive type as in many small insects which have no ventilatory movements. (2) A mixed type in insects, which ventilate their air sacs and the main tracheae whereas the exchange with the tissues occurs by diffusion via a tracheal system subdividing into small tracheolae which may indent the cells. (3) A convective type; the animals – some aerial fish, amphibians, reptiles, birds and mammals – ventilate their lungs with air; to this the classical equations for external respiration (see Rahn and Fenn, 1955) may generally be applied. However, it is true that some gas-phase diffusion process occurs deep in the lungs in the terminal bronchioles and in the alveolar sacs.

What about the aquatic insects? As is well known, there are two categories (Rahn and Paganelli, 1968): (1) Insects, such as *Dytiscus*, with a compressible gas pocket, or gas gill, into which the spiracles open. This gas pocket has a tendency to decrease and would eventually disappear if the insects did not surface from time to time to replenish it with fresh air. These aquatic insects are obviously air breathers, just as whales and seals are air breathers. (2) Insects, such as *Aphelocheirus*, with an incompressible gas gill or plastron. In these animals, a layer of gas, sequestered in hydrophobic hairs, covers some parts of the body to which the spiracles open. The volume of the plastron remains constant. Its nitrogen concentration is in equilibrium with the nitrogen dissolved in water and remains constant. Since the oxygen is used by the animal, its pressure in the plastron is lower than in the ambient well-aerated water, and O_2 diffuses from the water to the plastron's gas. Carbon dioxide diffuses out of the tracheae to the plastron and then dissolves in the water. The total pressure in the plastron is essentially independent of the insect's diving depth because the hydrostatic pressure is opposed by the tension of the air–water interface. If the hydrostatic pressure is higher, the radii of the contiguous menisci constituting the plastron decrease; consequently the tension developed according to the Laplace law counteracts the hydrostatic pressure. Finally the insects endowed with an incompressible gas gill are actually exchanging O_2 and CO_2 with a gas phase, which itself exchanges these gases with the surrounding water. The profiles of O_2 and CO_2 gradients along the gas-filled tracheae are exactly the same as in open-air insects. The insect is an air breather.

To help the understanding of this proposition, imagine the following scenario (whose feasibility was actually envisaged). Picture a human living in a large gas bubble deep in the water, made of a complex set of gas-permeable membranes interposed between the gas phase and the ambient water. The human would still ventilate and exchange O_2 and CO_2 with the gas of the bubble which he inhabits, this bubble getting its O_2 from the ambient water and dissolving its CO_2 into it. Although ultimately O_2 comes from the water and CO_2 goes into the water, the subject clearly remains an air breather. Only if he were able to take water into his lungs (a situation which has also been envisaged), would he be a water breather.

Skeleton and circulation

Animals emerging from water lose their buoyancy and are fully strained by gravity. Gravity affects two systems: the skeleton and the circulation.

Moreover, the larger the animal's size, the more stressful is the action of gravity. (Of course the vestibular apparatus is also affected by gravity; but, presumably, its sensitivity is not buoyancy-dependent.)

Skeleton. In agnaths and teleosts, the percentage of 'bone' of the total body mass is lower than in terrestrial animals and the proportion is constant whatever the body size. In birds and mammals the proportion of bone is higher, and increases with body size. For example, the bone mass is 4% of the whole body mass of a 4-g shrew and more than 25% of a 10-ton elephant. In contrast to terrestrial mammals, but like fishes, the percentage of skeletal mass in the wholly aquatic mammals, the whales, is almost independent of body size (Pace and Smith, 1981). For a 10-ton whale the skeletal mass is less than 15% of the total mass of the body.

The same allometric relations apply to other kinds of 'skeleton' or 'supportive tissues', namely the exoskeleton of spiders, and the shell of avian eggs (Anderson *et al.*, 1979). It would be premature, nonetheless, to draw a general conclusion as to the relative importance of skeleton in aquatic and terrestrial animals. For one thing, data are rare for diving mammals and for another, as far as I know, no difference of the relative shell mass between adult aquatic and terrestrial snails has yet been found.

Circulation. An aquatic animal's change of posture will not affect the blood pressure much, since the volumic masses of blood and water are similar. In terrestrial animals, on the other hand, the blood pressure depends on size and posture. In standing terrestrial vertebrates, the blood pressure is much higher in the lower than in the upper part of the body. In the standing giraffe, cow and man (Patterson *et al.*, 1965), the blood pressure in the vertebral and carotid arteries near the skull is relatively independent of the body posture, an observation which favors the concept of brain sparing (p. 109). The arterial blood pressure, much higher in the lower part of the body, may be 400 Torr in the leg arteries of the giraffe. The thickness of the arterial wall is related to the blood pressure; for example, in the giraffe, the posterior tibial artery wall is five times as thick as that of the carotid artery (Goetz and Keen, 1957).

More direct evidence of the role of hydrostatic pressure in the circulation of aquatic and terrestrial animals is offered by the study of arboreal, terrestrial, semi-aquatic and aquatic snakes (Seymour and Lillywhite, 1976). The heart is more posterior in aquatic snakes, 25–45% of the body length, than it is in terrestrial and arboreal species, 15–25%. If the sea snake is tilted in water, blood pressure does not change measurably. This is not surprising: because the specific masses of water and blood are similar, the circulation cannot change much. On the contrary, when sea snakes are tilted in air, head up, the cephalic blood pressure drops to zero and the animals lose consciousness. Arboreal and terrestrial snakes undergoing the same maneuver increase the aortic arch blood pressure and maintain their cephalic blood pressure, observations which speak in favor of the concept of brain sparing (p. 129). Here aquatic snakes have been compared to terrestrial snakes rather than the contrary, since the aquatic reptiles descend from terrestrial reptiles which have reinvaded water (Seymour, 1987).

Cost of breathing

The cost of breathing involves two factors: the mechanical work of breathing, and the efficiency of the respiratory muscles which ensure this work. For theoretical and experimental reasons, the evaluation of these two factors is difficult. Scheid (1987) recently presented a critical review of this problem.

The estimation of the first factor, namely the mechanical work rate (power) is very complex, because it includes many poorly quantified components: (1) the work to overcome the inertia of the inhaled medium, a work small in air breathers because of air's low density; (2) the work to overcome the inertia of the various anatomical elements of the ventilatory apparatus; (3) the work to change the volume and the shape of the more-or-less compliant structures of the breathing apparatus; (4) the work to overcome the resistance to flow of the respired medium. This resistance is proportional to the square of the flow rate for a laminar stream, a value which also depends on the milieu's density, and to the cube of the flow rate for a turbulent stream, which depends on the kinematic viscosity. Since the kinematic viscosity is higher in air than in water, the Reynolds number is lower. Air has a greater tendency to flow turbulently than water. On the other hand, because air has a relatively low density, little energy is required for it to be accelerated and decelerated and to flow laminarly. It is possible, finally, that air requires less mechanical power to flow than water.

These considerations concern the mechanical work. To know the real energy expense needed for breathing, *i.e.* the cost of breathing, one needs to know the efficiency of the activity of the respiratory muscles. This can be evaluated only if one knows first the overall energy required for breathing, a value necessary for comparing the cost of breathing in water- and air breathers. We then meet a circular argument.

It is very difficult to determine the cost of breathing directly. There have been various experimental protocols. The estimations of the cost of breathing water are extremely scattered, from a few percent to more than 50% of the overall oxygen consumption. A very interesting technique for estimating this cost is the measurement of the fall of oxygen consumption which occurs when the trout (Steffensen, 1985) or the sharksucker (Steffensen and Lomholt, 1983) passes from active ventilation to passive ventilation, *i.e.* ram ventilation*, when the velocity of water streaming over their body is adequate to ensure ventilation. Trout O_2 consumption falls by more than 10%, and the sharksucker's about 5%, in the passage from active to passive breathing. These values probably underestimate the actual cost of breathing, because active ventilation should decrease as the velocity of water increases, until the velocity is high enough to ensure a completely passive ventilation.

Steffensen (1985) has noted that the water velocity required for the transition of the resting sharksucker from active to ram ventilation increases with decreasing values of the water P_{O_2}. When the trout is passively ventilated, the work of breathing is made up by machinery which pumps water over the animal body. For fish which cruise open-mouthed without mouth-opercular movement, the cost of breathing is ensured by the swimming muscles, but the efficiency is better since the overall oxygen consumption decreases. Finally the sharksuckers' cost of breathing is paid by the shark to which they are attached.

Air breathers' cost of breathing is most probably less than that of water breathers. Apparently there is no study on birds. Kinney and White (1977) studied the cost of breathing of one reptile, the turtle *Pseudemys floridana*; it was high, increasing from 10% to 40% of the overall O_2 consumption when the body temperature decreased from 37 to 10 °C, but the animals had been made artificially hypocapnic, and hypocapnia may increase the resistance of the airways and have various metabolic effects.

---

* Roberts (1975) and Jones and Randall (1978) reported observations of ram ventilation made in several groups of fishes.

Furthermore chelonians may be particular since they breathe by moving their legs.

Among mammals, humans have usually been the subjects (Otis, 1964). The best methods are measurements of the increase of O_2 consumption resulting from an isocapnic voluntary hyperventilation (Cournand *et al.*, 1954) or of the hyperventilation resulting from the addition of instrumental dead spaces (Milic-Emili and Petit, 1960). The cost of breathing is low at rest, maybe 1–2% of the total O_2 consumption, but increases in hyperbary because the gas density increases, and in exercise, because work to overcome inertia and gas turbulence is increased.

In comparing the cost of breathing in water- and air breathers, the fact that for a given O_2 consumption air breathers breathe much less than water breathers may be the main reason why the relative cost of breathing may be less in air- than in water breathers. However, if the inhaled air is not water-saturated, and usually it is not, some water evaporates and consequently more energy is used. In the next section we will see that the complete homeothermy we observe in birds and in mammals could not have evolved if the specific ventilation and the pulmonary heat loss were not low, a situation which is possible, for a given O_2 tension, by the much higher O_2 concentration of air compared to water.

Heat loss and the development of homeothermy

Animal heat is transferred from the body to the environment in two stages: (1) heat is transferred from all parts of the body to the integument; (2) heat is transferred from the integument to the environment. Table 9.1 shows that heat is dissipated with much more difficulty in air than in water. Consequently it seems natural to infer that land animals were the first to become homeotherms. Some rare aquatic animals are homeothermic; they will be discussed specifically later on.

In water, heat transfer is so rapid that the skin temperature is very close to the ambient temperature, even in aquatic mammals with their relatively high metabolism. Furthermore, in all water breathing animals, *e.g.* molluscs, crustaceans, fish, the ventilated water carries away the heat brought to the gills by the venous blood. Post-branchial arterial blood is at the temperature of the ambient water.

In land animals, except most amphibians, and mammals which may sweat, the tegument is dry since air is rarely completely saturated with water vapor. If the skin is wet, as in the amphibians, its temperature is lower than the ambient temperature. The amphibians (with the exception of some treefrogs, p. 125) avoid desiccation by remaining in a wet atmosphere, or, if they venture into an unsaturated atmosphere, they return to water from time to time to restore their water content. Some water is evaporated via the ventilated air, but not much, except in special cases of increased ventilation which will be mentioned later.

Because of their low heat production, the body temperature of land reptiles is generally not much higher than the ambient temperature. However, some lizards may regulate their temperature behaviorally, since they may expose themselves to solar radiation to warm up, and move into the shade to cool down (*e.g.* see Schmidt-Nielsen, 1983, pp. 296–298). The Galapagos marine iguana (Bartholomew, 1987) adopts a similar strategy; it warms up in air by solar radiation, and during its dives for feeding the body temperature falls; on the return to land it warms up again. Some snakes, like the female coiled python brooding her eggs, may increase their heat production several times and have a body temperature a few degrees above ambient temperature (Hutchison *et al.*, 1966). Some insects, *e.g.* the bumblebee and the sphinx moth (Schmidt-Nielsen, 1983, pp. 299–301), may also have a temperature higher than the ambient air due to exposition to solar radiation and to muscular activity. When a high enough temperature is reached, they may fly and search for food. The scarab elephant beetle warms up without locomotion or any other overt activity when the ambient temperature falls below 20–22 °C (Morgan and Bartholomew, 1982); presumably a hormonal mechanism is involved. Finally, honeybees may regulate the temperature of their beehive at about 35 °C in summer and at 20–30 °C in the winter even if the ambient temperature falls well below 0 °C.

Birds and mammals differ from poikilotherms in that their energy metabolism is much more intense (at a comparable body mass) and their temperature constant for at least part of their lifetime. The origin of homeothermy is much disputed (Crompton *et al.*, 1978; McNab, 1978; Bennett and Ruben, 1986), but no one contests that it first occurred to its full extent in land animals. The first reason is that heat capacity and heat conductivity of air are much lower than those of water. Another reason, which is not paid much attention to, is that for a given O_2 consumption, air breathers breathe much less than water breathers. If air-breathing

TABLE 9.3

Oxygen consumption, ventilation, water loss and energy lost by ventilation in (1) normal human, and (2) a subject assumed to breathe air at the rate a water breather breathes water to obtain the same amount of oxygen.

Human	$\dot{M}_{O_2}$		$\dot{V}_E$	Loss of water by ventilation	Loss of energy by water vaporization	
	$mmol \cdot min^{-1}$	W	$ml \cdot min^{-1}$	$g \cdot min^{-1}$	$J \cdot min^{-1}$	W
Normal	14	105	10	0.22	530	9
Breathing air at the rate of a water breather*	14	105	250	5.5	13 250	220

The energy equivalent of oxygen is 450 $J \cdot mmol^{-1}$. * The increase of $\dot{M}_{O_2}$ due to the increased ventilation is neglected.

vertebrates ventilated as much air as the fish ventilates water, they would not have been able to become homeothermic, and at a temperature which may be very much higher than the ambient temperature. This is not so much because they could have lost heat by warming the air they breathe, since the heat capacity of air is relatively low, but they would have lost considerable heat by evaporating water. Indeed, exhaled gas is water-saturated whereas inhaled gas practically never is.

An example will show how much energy is saved by the relatively low ventilation of air breathers, low as compared to water breathers for the same oxygen consumption. We will take an example in human physiology since the human has no respiratory peculiarities compared to other mammals (table 9.3).

A resting human may breathe at sea level an average of 10 $L \cdot min^{-1}$ in order to get 14 mmol $O_2 \cdot min^{-1}$. If the inspired air at 20 °C is 50% water saturated and the expired gas at 32 °C is 100% saturated, the subject loses 0.22 g of water per minute (about 350 g per day). The vaporization of this water has a cooling effect which is the product of the mass of vaporized water times the specific heat of vaporization, that is 530 $J \cdot min^{-1}$ (*i.e.* about 9 W).

Let us suppose that the human had to breathe as much air as a water breather has to breathe water to get 14 mmol $O_2 \cdot min^{-1}$. Then the human would ventilate 250 L $\cdot$ min^{-1}, *i.e.* 25 times more than his actual ventilation. Consequently the water loss and the energy loss by water evaporation would be respectively 5.5 g $\cdot$ min^{-1} and 13 250 J $\cdot$ min^{-1} (220 W). This is obviously impossible for various reasons: (1) no human can ventilate at more than 200 L $\cdot$ min^{-1} even for a very short time; (2) the loss of water would be enormous, amounting to about 8 L per day; (3) the rate of energy loss by water cooling would be very high, 220 W. Furthermore, our example was of a resting man; if the subject is reasonably active his oxygen consumption, his ventilation and consequently his rate of water evaporation and of cooling are increased by a factor of 2–3. In these calculations, there are some assumptions; in particular the heat necessary to warm the 20 °C inspired air to the 32 °C expired gas is neglected.

We come to the conclusion that it is impossible to imagine full homeothermy developing in water breathers, or in terrestrial air breathers breathing at the rate of a water breather, since the energy lost by cooling (220 W) would be higher than the energy produced by metabolism (105 W). It cannot be argued that the animal could consume more oxygen to cover this heat loss, because any increase of oxygen consumption would necessarily augment the ventilation *and* the heat loss. One may note in passing that humans exposed to very stressful ambient heat or radiation can have such an energy loss by water vaporization, but in this case the heat gained from outside and the metabolic heat may be balanced by the cooling effect of sweat vaporization.

This counter-example shows that general homeothermy could not develop in an air breather breathing at the rate of a water breather. If the O_2 capacitance in air were the same as in water, then some animals could invade aerial niches which would have to be water-saturated to obviate desiccation. They would also have to be close to a source of water for the wash-out of ammonia, unless they had developed the means to avoid ammonia toxicity, by excreting urea or uric acid. In fact, some amphibians which depend to some degree on their cutaneous respiration for gas exchange live in damp places, near water. It is not conceivable that such animals could become endothermic homeotherms.

To colonize open land, animals must (1) excrete an appropriate nitrogenous end product, (2) save water by eliminating organic and mineral substances at high concentrations, and (3) lower the ventilation to save water and to avoid the cooling effect of water vaporization.

It is because the capacitance of air for oxygen is high, compared to its value in water, that the ventilation could be lowered, resulting in a decrease of the water and energy losses. Such a lowering of ventilation entails an increase of the pulmonary CO_2 partial pressure, a hypercapnia. We have seen (p. 41) that $[HCO_3^-]$ in air breathers is increased proportionally to the rise of P_{CO_2}, so that the pH values, at the same temperature, are similar. The important increase of the bicarbonate concentration raises a problem of ionoregulation of the body fluids, as discussed above (p. 127).

Land colonizers so prepared could become homeothermic. Homeothermy requires an intense metabolism (Bennett and Ruben, 1979; Karasov and Diamond, 1985), heat insulation of the body, with the possibility of regulating this insulation and, eventually, of cooling by water evaporation. Indeed, homeothermy means that the temperature is maintained steady against the variations of the environmental temperature and of all the factors, humidity, radiation, convection, conduction, which affect heat exchanges between the body and its surrounding, and also against the increases of heat production caused by exercise and nutrition. Generally the body temperature is higher than the ambient temperature, and to avoid a great loss of heat, the animal must insulate its body, as it does by a thick epidermis, by feathers, hair and fur.

But eventually, in order to meet the physiological increase of heat production and the natural variations of the environment which hinder heat loss or, even in some cases, heat the body, homeothermic animals must also dispose of an extra mechanism of cooling, namely an appropriate water evaporation. Some animals sweat. Others which do not possess sweat glands use their ventilation to evaporate water and to cool, by two non-exclusive mechanisms: gular flutter in some birds and panting in some birds and some mammals, which increase ventilation and water evaporation (p. 101). But what about that condition which permitted the colonization of open land and the emergence of homeothermy, namely the decrease of ventilation which resulted in a hypercapnia? Now we invoke an increase of ventilation as a means of keeping the body temperature steady. There is no contradiction here, because the hyperventilation, gular flutter or panting, observed during environmental thermal stress is of a special type: it is a superficial ventilation with a small tidal volume and a very high frequency; consequently the ventilation of the lung with fresh air is not increased (p. 102).

We conclude that homeothermy is characteristic of land animals. Immediately two criticisms come to mind: (1) imperfect homeothermy exists in some fishes, the warm-bodied fish, which are neither terrestrial, nor air breathing; (2) full homeothermy exists in aquatic mammals, such as the whales. Let us consider these two cases.

(1) *Warm-bodied fishes*. Indeed, in many wholly aquatic animals (aquatic in all respects, including water breathing), the temperature of some parts of the body is higher than the ambient temperature (*e.g.* see Graham, 1975; Carey, 1982). Many tuna and sharks are warm-bodied: in these animals, some muscles, the eyes and the brain, and the viscera during digestion, are warmer than the ambient water. The hyperthermic level, the difference between the local and ambient temperature, may reach a few degrees. Only in the bluefin tuna is there clear evidence that the temperature of certain swimming muscles can be much higher than the ambient temperature: Carey and Teal (1969) observed that the warmest portion of the muscle mass varies only between 27 and 32 °C whereas the ambient temperature varies between 7 and 30 °C. Analogous observations for the brain and the eyes were made in the bluefin tuna by Linthicum and Carey (1972). This demonstrates that in the bluefin tuna, there exists some imperfect thermoregulation (Carey and Teal, 1969; Carey, 1982). In other warm-bodied fish the regulation, if any, is very imperfect.

How do warm-bodied fish conserve heat? There are two mechanisms. (1) A countercurrent heat exchanger with a *rete mirabilis* creates a kind of thermic barrier of the organ by heat recirculation. (2) Some special heating tissues, such as modified eye muscles, heat the brain and the eyes; they have been described in fish such as marlin and sailfish which are otherwise cold-bodied (Block, 1987).

In conclusion, one may say that some wholly aquatic fish possess imperfect thermoregulation, generally localized to some organ. The bluefin tuna is, as far as is known, the most successful fish in this respect; it is remarkable for an animal which lives in a highly heat-dissipative milieu, water, in which it may cruise at eight knots.

(2) *Aquatic mammals*. They are the descendants of terrestrial mammals which have recolonized water, and, like terrestrial mammals, they are ureotelic, air-breathing and they reproduce like other mammals (eutherians). They obviously have some piscine traits of locomotion. However, these mammals are homeothermic in spite of their aquatic,

often very cold habitats. Being air breathers they breathe little as compared to water breathers, and consequently have a low respiratory heat loss. Their homeothermy is limited to the deep core of their body, insulated from the cold water by thick blubber which can reach 50% of the whole body mass. There are no lean whales. During prolonged dives, the temperature of some organs, like the kidneys and the gastro-intestinal tract whose steady function is not vital, may drop markedly as a result of vasoconstriction, and the active circulation is limited to the heart, the brain and the lung. We have already dealt with this phenomenon in the section on brain sparing (p. 109). When whales or seals surface, the circulation to all parts of the body is restored and the oxygen debt, originating in the depletion of the oxygen stores and the accumulation of fermentation products, mainly lactic acid, is paid back. Certainly these phenomena are observed only during deep and prolonged dives, but they are vital for the survival of the species. Routine dives are short and aerobic without curtailment of the circulation in any organ (p. 94).

Homeothermy in aquatic mammals is reduced to the core of the body because of the thickness of blubber. However, in no mammal or bird is the temperature the same throughout the whole body. The skin temperature is always below the core temperature, creating a gradient of temperature which allows a flux of heat. The core temperature itself is poorly defined; the rectal, or better the colonic, temperatures are generally taken as good estimates of the core temperature. Probably the cardiac temperature is the best core temperature, since it is there that the blood of all parts of the body is mixed. To sum up, what is called the temperature of a homeotherm, *e.g.* 37 °C, should be understood as the core temperature; the *average* body temperature is several degrees lower. Bazett *et al.* (1948) have drawn attention 'to the chaos introduced into physiology by the fictitious assumption of a constant blood temperature'.

Life in water and on land

Starting from respiratory physiology which shows marked contrasts between water- and air breathers, obviously in relation with the special characteristics of the two milieus with respect to O_2 and CO_2, can other physiological traits in aquatic and terrestrial animals also be contrasted? Contrasts as to nitrogenous end products, water conservation, concentration of excreted organic molecules and salts, and body temperature have been discussed. We dealt with these functions because they are immediately related to respiration, but it is well known that locomotion, sensory mechanisms and reproduction are very different in terrestrial and aquatic animals.

Table 9.4 is a schematic resume of the main differences in wholly aquatic and wholly terrestrial groups. They are observed in many groups

TABLE 9.4
Some physiological characteristics of typical aquatic and terrestrial animals.

	Aquatic	Terrestrial
Respiration	skin and/or gills	tracheae or lungs
$\Delta P_{CO_2}/\Delta P_{O_2}$	low ratio	high ratio
$[HCO_3^-]$	low	high
N end products	mainly ammonia	mainly urea and/or uric acid and other purine derivatives
Water turnover	high or very high	low or very low
Locomotion	swimming	running, flying
Temperature	poikilothermy (most of them)	homeothermy (some of them)

ΔP_{O_2} and ΔP_{CO_2} designate the difference of P_{O_2} and P_{CO_2} values between body fluids and ambient milieu.

TABLE 9.5
Main animal groups with aquatic and terrestrial forms.

Phylum	Aquatic	Terrestrial
Nematodes	+	few
Annelids	+	+
Molluscs	+	+ (only snails)
Arthropods	+	+ {Arachnids, Myriapods, Insects}
Vertebrates	+	+

(table 9.5) which are unconnected by evolution; this fact demonstrates that some *convergences* of functions are related to the characteristics of the milieus.

The broad division of the animal kingdom into aquatic and terrestrial animals is very old. It was a distinction made by anatomists and zoologists who in the past inferred the nature of functions from the knowledge of the structures. Later, with the development of physiology and biochemistry, more contrasts between aquatic and terrestrial life became evident (Rahn, 1967; Dejours, 1979a; Wald, 1981; Schmidt-Nielsen, 1983; Little, 1983; Comparative Physiology: Life in Water and on Land, 1987).

Many books of comparative physiology do not attempt such a distinction. One reason may be that the contrasts between terrestrial and aquatic animals may not be general. Whereas in some terrestrial animals, *e.g.* insects, all functions are different from those of aquatic animals, *e.g.* aquatic crustaceans, some animals are intermediate: (1) Some are amphibious, because they spend a part of their lifetime in water and the rest in air, as intertidal animals or some amphibians. (2) Some nematodes, annelids and arthropods are intermediate because they live in a more-or-less wet soil. (3) Other animals such as turtles and cetaceans are descendants of terrestrial animals, and they keep some traits of their terrestrial origin. For example, whales and porpoises are air breathers and have the respiratory and excretory characteristics of their ancestors, but because their gross body shape is fishlike, the cetaceans were classified until the 18th century among the Pisces. Animals like the aquatic insects, reptiles and mammals are obviously aquatic by their abode, but physiologically they are intermediate; certain functions – locomotion, circulation, sensory mechanisms – are of an aquatic nature; some others – such as respiration, nitrogen metabolism – are terrestrial.

The existence of intermediate animals should not restrain us from contrasting completely terrestrial and completely aquatic animals, for two reasons:

(1) All animal classifications meet similar problems of borderline, intermediate animals. Indeed we easily contrast:

– Invertebrates and vertebrates, but we have the non-vertebrate chordates.

– Osmoconformers and osmoregulators, but many animals do not enter into either category and may be considered as imperfect osmoregulators.

– Poikilotherms and homeotherms, but, as we have seen, some poikilotherms have features of homeotherms (*e.g.*, some social insects, and some tunas) and in some higher vertebrates homeothermy may be imperfect or intermittent.

(2) On the contrary, the existence of intermediate animals may be instructive because their study may shed light upon the process by which some aquatic animals could invade land, and particularly how some terrestrial animals could become homeothermic.

Conclusion

Water and air become inhospitable milieus at low and high temperature. Cold may endanger life of some animals above the freezing point, *e.g.* the human heart will stop beating at a body temperature of some 20–25 °C. At freezing temperatures, the body waters may freeze; but some fish, some invertebrates, mainly insects, can supercool, *i.e.* can keep a body temperature below the freezing temperature without crystal formation thanks to the presence of some antifreezes, such as glycerol (*e.g.* see Hardy, 1972).

High temperature endangers life because it may denature proteins or may disturb the function of some organs, such as the brain, or the cardiovascular system (collapsus). The increase of temperature has two kinds of action: (1) on the environment, and (2) on the metabolism.

(1) Figure 9.1 shows that the solubilities of O_2, CO_2 and NH_3 in water (here distilled water) decrease markedly with a rise of temperature. For CO_2, if the water contains some buffers, the action of temperature may be complex. Also the case of ammonia is not simple because of the interaction between the carbonic and the ammoniacal systems. However, it remains that the higher the water temperature, the lower is its capacity to fix CO_2 and NH_3 and to deliver oxygen.

(2) On the other hand, fig. 9.1 shows in poikilotherms the effect of temperature on the metabolism rates $\dot{M}_x$, x standing for O_2, CO_2, proteins and nucleoproteins, electrolytes, since with an increase of temperature the rate of energy expenditure augments as well the food intake and the turnover of electrolytes and organic substances, the O_2 consumption and the CO_2 production. The line 'metabolisms' of fig. 9.1 is drawn assuming a constant Q_{10} of 2.5. In reality the value of Q_{10} may differ from 2.5 and may not be constant; but these restrictions do not alter the point of the present argument.

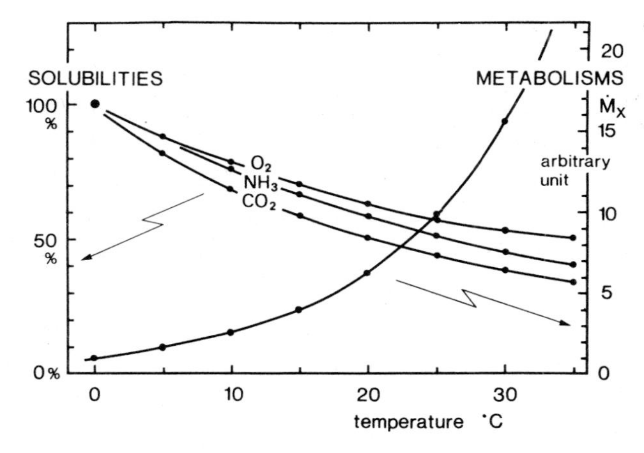

Fig. 9.1. Metabolic rates of a poikilothermic water breather and relative solubilities of O_2, CO_2 and NH_3 in distilled water as a function of temperature. The left ordinate scale concerns the relative solubilities of O_2, CO_2 and NH_3 when one takes the actual values of solubilities at 0 °C as 100%. For example, at 35 °C the oxygen solubility is half its value at 0 °C. The right ordinate scale concerns the rates of metabolisms, *i.e.* O_2 consumption, CO_2 production, nitrogenous catabolism. The unit of this scale is taken arbitrarily as 1 for the value at 0 °C and has no dimension. All metabolisms are assumed to follow the same thermal increment of intensity with a constant value of Q_{10} at 2.5.

That is, water breathers at high temperature have a high metabolism, need more oxygen, produce more carbon dioxide and more nitrogenous end-products, whereas the water they live in contains less oxygen, is often hypercapnic (Ultsch, 1987) and does not accept metabolic wastes as easily as cold water. For example, if a poikilotherm is switched from 0 to 35 °C, its O_2 consumption may be increased 25 times whereas the ambient O_2 concentration for a given P_{O_2} value is halved. Obviously high temperature creates an *ecological pressure* for water breathers. The ecological pressure could still be more important if the first amphibious dual breathers were marine or estuarine fish, since O_2 and CO_2 solubilities are lower in salt water than in freshwater (Packard, 1974), a view which is debated by Graham *et al.* (1978). One might find support for this concept of ecological pressure in which high O_2 consumption and low O_2 pressure in warm water play an important role in the emergence of air breathing, in the observation that the experimental increase of O_2 pressure in water prolongs the time of survival in goldfish placed at a lethal temperature or raises the lethal temperature which they reach as a result of heating (Weatherley, 1970).

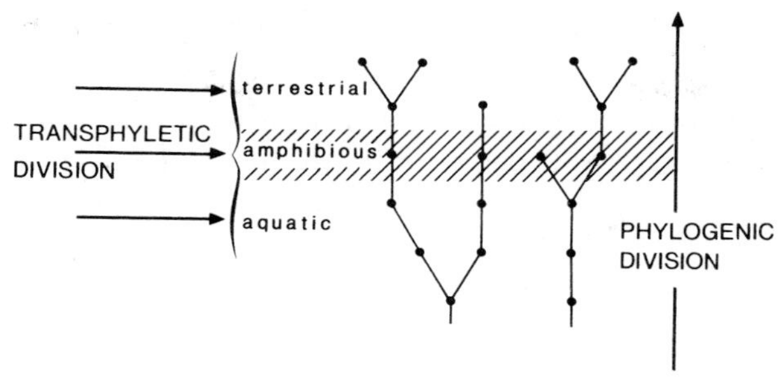

Fig. 9.2. Classical phylogenic tree to which is superimposed a transphyletic division, based on the nature of the milieu which entails special designs and functions in the animal groups.

In natural conditions, animals, if they possess the right apparatus for gas exchange, can fix more oxygen by using air as a respiratory milieu, a milieu which accepts carbon dioxide as easily as water does. But to be permanently terrestrial and to invade dry areas, the animals must have solved the problem of ammonia toxicity, by forming some little toxic N-containing derivatives, as urea and uric acid. However, an air breather which is still ammoniotelic can use the high availability of oxygen in the atmosphere as long as it is able to wash out its ammonia from time to time by returning to water, which is an almost infinite sink for this very soluble and diffusible substance. Presumably this amphibious life was a necessary stage in the evolution from aquatic to terrestrial arthropods and from fish to terrestrial vertebrates (fig. 9.2).

To sum up, whether aquatic animals are invertebrates or vertebrates, the problems raised by the invasion of land (non-toxic nitrogen end products; water conservation; concentrated excretions of salts and some organic compounds; relatively low specific ventilation, hypercapnia and the ensuing changes of acid–base balance; compensatory mechanisms for the action of gravity on circulation and on skeletons; special designs for locomotion and for communications; internal fecundation) are common to all. Many physiological traits, although supported by quite different biochemical and morphological structures, are convergent and related to the environmental characteristics.

Thus one is led to suggest an ecophysiological transphyletic division of the animal kingdom which is generally not discussed by evolution biologists (fig. 9.2). This division is an example of so-called 'artificial' classifications, more precisely of teleological classifications. But within the phyla, molluscs, arthropods, vertebrates, the classical phylogenic approach to problems of evolution has obviously to take into account the nature of the milieu. The different approaches of 'evolution and milieu' are probably complementary.

Appendix

TABLE A.1
Saturation pressure of water vapor at various temperatures T′.

T′ (°C)	Pressure (Torr)	T′ (°C)	Pressure (Torr)	T′ (°C)	Pressure (Torr)	T′ (°C)	Pressure (Torr)
− 1	4.3						
0	4.6	10	9.2	20	17.5	30	31.8
1	4.9	11	9.8	21	18.7	31	33.7
2	5.3	12	10.5	22	19.8	32	35.7
3	5.7	13	11.2	23	21.1	33	37.7
4	6.1	14	12.0	24	22.4	34	39.9
5	6.5	15	12.8	25	23.8	35	42.2
6	7.0	16	13.6	26	25.2	36	44.6
7	7.5	17	14.5	27	26.7	37	47.1
8	8.0	18	15.5	28	28.3	38	49.7
9	8.6	19	16.5	29	30.0	39	52.4
						40	55.3
						41	58.3
						42	61.5

TABLE A.2
Values of the thermodynamic constant R of ideal gas, for different units of pressure, of volume, of energy (= pressure × volume), of quantity of substance. All units are expressed per degree Kelvin, K.

Pressure	Volume	Energy	Quantity of substance	R value	Unit
Pa	m^3	J	mol	8.314	$J \cdot (K \cdot mol)^{-1}$
atm	L	atm · L	mol	0.082 05	$atm \cdot L \cdot (K \cdot mol)^{-1}$
psi	L	psi · L	mol	1.206	$psi \cdot L \cdot (K \cdot mol)^{-1}$
Torr	L	Torr · L	mol	62.4	$Torr \cdot L \cdot (K \cdot mol)^{-1}$
Torr	L	Torr · L	LSTPD	2.785	$Torr \cdot L \cdot (K \cdot LSTPD)^{-1}$
		cal	mol	1.987	$cal \cdot (K \cdot mol)^{-1}$

TABLE A.3

O_2 and CO_2 capacitances, βw_{O_2} and βw_{CO_2}, in distilled water, sea water (salinity 35.1‰; chlorinity 19.4‰) and in air at various temperatures T′, in °C. The unit is $\mu mol \cdot L^{-1} \cdot Torr^{-1}$.

T′	Distilled water		Sea water		Air
	βw_{O_2}	βw_{CO_2}	βw_{O_2}	βw_{CO_2}	βg
43	1.3144	29.317	1.1026	25.909	50.69
42	1.3279	29.915	1.1127	26.382	50.85
41	1.3419	30.537	1.1231	26.874	51.01
40	1.3567	31.182	1.1340	27.386	51.17
39	1.3721	31.853	1.1454	27.918	51.34
38	1.3882	32.550	1.1574	28.473	51.50
37	1.4050	33.274	1.1699	29.050	51.67
36	1.4225	34.028	1.1829	29.652	51.84
35	1.4408	34.812	1.1965	30.278	52.00
34	1.4598	35.627	1.2107	30.930	52.17
33	1.4797	36.476	1.2255	31.610	52.34
32	1.5003	37.359	1.2409	32.319	52.52
31	1.5218	38.280	1.2570	33.058	52.69
30	1.5441	39.238	1.2737	33.828	52.86
29	1.5673	40.237	1.2910	34.632	53.04
28	1.5915	41.279	1.3091	35.471	53.21
27	1.6167	42.365	1.3279	36.347	53.39
26	1.6428	43.498	1.3475	37.261	53.57
25	1.6700	44.681	1.3678	38.216	53.75
24	1.6982	45.915	1.3889	39.214	53.93
23	1.7276	47.204	1.4109	40.257	54.11
22	1.7582	48.551	1.4338	41.347	54.29
21	1.7900	49.959	1.4575	42.487	54.48
20	1.8231	51.431	1.4823	43.681	54.67
19	1.8576	52.970	1.5080	44.930	54.85
18	1.8934	54.581	1.5347	46.237	55.04
17	1.9308	56.267	1.5626	47.607	55.23
16	1.9696	58.034	1.5916	49.043	55.42
15	2.0101	59.884	1.6218	50.548	55.61
14	2.0523	61.824	1.6532	52.126	55.81
13	2.0963	63.857	1.6860	53.782	56.00
12	2.1422	65.991	1.7202	55.521	56.20
11	2.1900	68.231	1.7558	57.347	56.40
10	2.2399	70.583	1.7929	59.265	56.60
9	2.2921	73.054	1.8316	61.282	56.80
8	2.3465	75.651	1.8720	63.403	57.00
7	2.4034	78.382	1.9142	65.634	57.20
6	2.4628	81.256	1.9583	67.983	57.41
5	2.5250	84.281	2.0044	70.457	57.61
4	2.5900	87.467	2.0525	73.064	57.82
3	2.6581	90.824	2.1029	75.813	58.03
2	2.7294	94.363	2.1556	78.712	58.24
1	2.8041	98.097	2.2108	81.772	58.45
0	2.8824	102.04	2.2686	85.004	58.67
−1	2.9646	106.20	2.3292	88.418	58.88
−2	3.0509	110.60	2.3928	92.028	59.10

References

Adams, W.E. (1958). The Comparative Morphology of the Carotid Body and Carotid Sinus. Springfield, IL, C.C. Thomas, 272 p.

Adolph, E.F. (1969). Regulations during survival without oxygen in infant mammals. *Respir. Physiol.* 7: 356–368.

Adolph, E.F. (1979). Development of dependence on oxygen in embryo salamanders. *Am. J. Physiol.* 236: R282–R291.

Anderson, J.F., H. Rahn and H.D. Prange (1979). Scaling of supportive tissue mass. *Q. Rev. Physiol.* 54: 139–148.

Angersbach, D. and H. Decker (1978). Oxygen transport in crayfish blood: effect of thermal acclimation, and short-term fluctuations related to ventilation and cardiac performance. *J. Comp. Physiol.* 123B: 105–112.

Anthonisen, N.R., D. Bartlett and S.M. Tenney (1965). Postural effect on ventilatory control. *J. Appl. Physiol.* 20: 191–196.

Ar, A., R. Arieli and A. Shkolnik (1977). Blood–gas properties and functions in the fossorial mole rat under normal and hypoxic–hypercapnic atmospheric conditions. *Respir. Physiol.* 30: 201–219.

Arieff, A.I., A. Kerian, S.G. Massry and J. DeLima (1976). Intracellular pH of brain: alterations in acute respiratory acidosis and alkalosis. *Am. J. Physiol.* 230: 804–812.

Arieli, R. (1979). The atmospheric environment of the fossorial mole rat (*Spalax ehrenbergi*): effects of season, soil texture, rain, temperature and activity. *Comp. Biochem. Physiol.* 63A: 569–575.

Arp, A.J. and J.J. Childress (1981). Functional characteristics of the blood of the deep-sea hydrothermal vent brachyuran crab. *Science* 214: 559–561.

Asmussen, E. (1983). Control of ventilation in exercise. In: Exercise and Sport Sciences Reviews, edited by R.L. Terjung. Am. College of Sports Medicine Series, pp. 24–54.

Åstrand, P.O. and K. Rodahl (1977). Textbook of Work Physiology. New York, McGraw-Hill Book Company, 681 p.

Badeer, H.S. (1986). Does gravitational pressure of blood hinder flow to the brain of the giraffe? *Comp. Biochem. Physiol.* 83A: 207–211.

Baker, M.A. (1982). Brain cooling in endotherms in heat and exercise. *Annu. Rev. Physiol.* 44: 85–96.

Ballintijn, C.M. (1982). Neural control of respiration in fishes and mammals. In: Proc. Third Congress of ESCPB. Vol. 1, edited by A.D.F. Adink and N. Spronk. Oxford, Pergamon Press, pp. 127–140.

Band, D.M., R.A.F. Linton, R. Kent and F.L. Kurer (1985). The effect of peripheral chemodenervation on the ventilatory response to potassium. *Respir. Physiol.* 60: 217–225.

Barcroft, J. (1938). The Brain and its Environment. New Haven, Yale University Press, 117 p.

Barnhart, M.C. (1986). Respiratory gas tensions and gas exchange in active and dormant land snails, *Otala lactea. Physiol. Zool.* 59: 733–745.

Barnhart, M.C. and B.R. McMahon (1987). Discontinuous carbon dioxide release and metabolic depression in dormant land snail. *J. Exp. Biol.* 128: 123–138.

Bartholomew, G.A., R.C. Lasiewski and E.C. Crawford (1968). Patterns of panting and gular flutter in cormorants, pelicans, owls and doves. *The Condor* 70: 31–34.

Bartholomew, G.A. (1987). Living in two worlds: the marine iguana, *Amblyrhynchus cristatus.* In: Comparative Physiology: Life in Water and on Land. Fidia Research Series, Vol. 9, edited by P. Dejours, L. Bolis, C.R. Taylor and E. Weibel. Padova, Liviana Press, distributed by Springer-Verlag, New York, pp. 389–400.

Baumgarten-Schumann, D. and J. Piiper (1968). Gas exchange in the gills of resting unanesthetized dogfish (*Scyliorhinus stellaris*). *Respir. Physiol.* 5: 317–325.

Bazett, H.C., L. Love, M. Newton, L. Eisenberg, R. Day and R. Forster (1948). Temperature changes in blood flowing in arteries and veins in man. *J. Appl. Physiol.* 1: 3–19.

Beaver, W.L., K. Wasserman and B.J. Whipp (1973). On-line computer analysis and breath-by-breath graphical display of exercise function tests. *J. Appl. Physiol.* 34: 128–132.

Beaver, W.J., N. Lamarra and K. Wasserman (1981). Breath-by-breath measurement of true alveolar gas exchange. *J. Appl. Physiol.* 51: 1662–1975.

Behrman, R.E., M.H. Lees, E.N. Peterson, C.W. De Lannoy and A.E. Seeds (1970). Distribution of the circulation in the normal and asphyxiated fetal primate. *Am. J. Obst. Gynecol.* 108: 956–969.

Benchetrit, G., J. Armand and P. Dejours (1977). Ventilatory chemoreflex drive in the tortoise, *Testudo horsfieldi. Respir. Physiol.* 31: 183–191.

Benchetrit, G. and P. Dejours (1980). Ventilatory CO_2 drive in the tortoise *Testudo horsfieldi. J. Exp. Biol.* 87: 229–236.

Bennett, A.F. and J.A. Ruben (1979). Endothermy and activity in vertebrates. *Science* 206: 649–654.

Bennett, A.F. and J.A. Ruben (1986). The metabolic and thermoregulatory status of therapsids. In: The Ecology and Biology of Mammalian-Like Reptiles, edited by N. Hotton, P.D. MacLean, J.J. Roth and C.E. Roth. Washington, D.C., Smithsonian Institution Press, pp. 207–218.

Bentley, P.J. (1966). Adaptations of amphibia to arid environments. *Science* 152: 619–623.

Bert, P. (1870). Leçons sur la Physiologie Comparée de la Respiration. Paris, Baillière, 588 p.

Biscoe, T.J. (1971). Carotid body: Structure and function. *Physiol. Rev.* 51: 437–495.

Bisgard, G.E. and J.H.K. Vogel (1971). Hypoventilation and pulmonary hypertension in calves after carotid body excision. *J. Appl. Physiol.* 31: 431–437.

Bjurstedt, H., C.M. Hesser, G. Liljestrand and G. Matell (1962). Effects of posture on alveolar–arterial CO_2 and O_2 differences and on alveolar dead space in man. *Acta Physiol. Scand.* 54: 65–82.

Bjurstedt, H., G. Rosenhamer and O. Wigertz (1968). High-G environment and responses to graded exercise. *J. Appl. Physiol.* 25: 713–719.

Black, C.P. and S.M. Tenney (1980). Oxygen transport during progressive hypoxia in high-altitude and sea-level waterfowl. *Respir. Physiol.* 39: 217–239.

Bligh, J., J.L. Cloudsey-Thompson and A.G. Macdonald (1976). Environmental Physiology of Animals. Oxford, Blackwell Scientific Publ., 456 p.

Bliss, D.E. and L.H. Mantel (1968). Adaptations of crustaceans to land: A summary and analysis of new findings. *Am. Zool.* 8: 673–685.

Bliss, D.E. (1979). From sea to tree: saga of a land crab. *Am. Zool.* 19: 385–410.

Block, B.A. (1987). Strategies for regulating brain and eye temperatures: a thermogenic tissue in fish. In: Comparative Physiology: Life in Water and on Land. Fidia Research Series, Vol. 9, edited by P. Dejours, L. Bolis, C.R. Taylor and E. Weibel. Padova, Liviana Press, distributed by Springer-Verlag, New York, pp. 401–420.

Boggs, D.F., D.L. Kilgore and G.F. Birchard (1984). Respiratory physiology of burrowing mammals and birds. *Comp. Biochem. Physiol.* 77A: 1–7.

Boutellier, U., R. Arieli and L.E. Farhi (1985). Ventilation and CO_2 response during +Gz acceleration. *Respir. Physiol.* 62: 141–151.

Boutellier, U. and L.E. Farhi (1986). Influence of breathing frequency and tidal volume on cardiac output. *Respir. Physiol.* 66: 123–133.

Boutillier, R.G. and D.P. Toews (1981). Respiratory, circulatory and acid–base adjustments to hypercapnia in a strictly aquatic and predominantly skin-breathing urodele, *Cryptobranchus alleganiensis*. *Respir. Physiol.* 46: 177–192.

Bouverot, P. and L.M. Leitner (1972). Arterial chemoreceptors in the domestic fowl. *Respir. Physiol.* 15: 310–320.

Bouverot, P., V. Candas and J.P. Libert (1973). Role of the arterial chemoreceptors in ventilatory adaptation to hypoxia of awake dogs and rabbits. *Respir. Physiol.* 17: 209–219.

Bouverot, P., G. Hildwein and G. Le Goff (1974). Evaporative water loss, respiratory pattern, gas exchange and acid–base balance during thermal panting in Pekin ducks exposed to moderate heat. *Respir. Physiol.* 21: 255–269.

Bouverot, P. (1978). Control of breathing in birds compared with mammals. *Physiol. Rev.* 58: 604–655.

Bouverot, P., D. Douguet and P. Sébert (1979). Role of the arterial chemoreceptors in ventilatory and circulatory adjustments to hypoxia in awake Pekin ducks. *J. Comp. Physiol.* 133B: 177–186.

Bouverot, P. (1985). Adaptation to Altitude–Hypoxia in Vertebrates. Berlin, Heidelberg, Springer-Verlag, 176 p.

Bradbury, M. (1979). The Concept of a Blood–Brain Barrier. New York, John Wiley and Sons, 454 p.

Bradford, S.M. and A.C. Taylor (1982). The respiration of *Cancer pagurus* under normoxic and hypoxic conditions. *J. Exp. Biol.* 97: 273–288.

Bramble, D.M. and D.R. Carrier (1983). Running and breathing in mammals. *Science* 219: 251–256.

Bridges, C.R. and P. Scheid (1982). Buffering and CO_2 dissociation of body fluids in the pupa of the silkworm moth, *Hyalophora cecropia*. *Respir. Physiol.* 48: 183–197.

Broyles, R.H. (1981). Changes in the blood during amphibian metamorphosis. In: Metamorphosis. A Problem in Developmental Biology. 2nd edition, edited by L.I. Gilbert and E. Frieden. New York, Plenum Press, pp. 461–490.

Bureau International des Poids et Mesures (1985). The International System of Units. 5th edition. Sèvres, France, Pavillon de Breteuil, 110 p.

Burggren, W.W. and B.R. McMahon (1981). Hemolymph oxygen transport, acid–base status and hydromineral regulation during dehydration in three terrestrial crabs, *Cardiosoma, Birgus* and *Coenobita*. *J. Exp. Zool.* 218: 53–64.

Burkett, B.N. and H.A. Schneiderman (1974). Roles of oxygen and carbon dioxide in the control of spiracular function in *Cecropia pupae*. *Biol. Bull.* 147: 274–293.

Burnett, L.E., T.N. Dunn and R.L. Infantino (1985). The function of carbonic anhydrase in crustacean gills. In: Transport Processes, Iono- and Osmoregulation, edited by R. Gilles and M. Gilles-Baillien. Berlin, Heidelberg, Springer-Verlag, pp. 159–168.

Burnett, L.E. and B.R. McMahon (1985). Facilitation of CO_2 excretion by carbonic anhydrase located on the surface of the basal membrane of crab gill epithelium. *Respir. Physiol.* 62: 341–348.

Burnett, L.E. and B.R. McMahon (1987). Gas exchange, hemolymph acid–base status, and the role of branchial water stores during air exposure in three littoral crab species. *Physiol. Zool.* 60: 27–36.

Burtin, B., J.C. Massabuau and P. Dejours (1986). Ventilatory regulation of extracellular pH in crayfish exposed to changes in water titration alkalinity and NaCl concentration. *Respir. Physiol.* 65: 235–243.

Busija, D.W. and D.D. Heistad (1984). Factors involved in the physiological regulation of the cerebral circulation. *Rev. Physiol. Biochem. Pharmacol.* 101: 161–211.

Cabanac, M. and M. Caputa (1979). Natural selective cooling of the human brain: evidence of its occurrence and magnitude. *J. Physiol. (London)* 286: 255–264.

Calder, W.A. and K. Schmidt-Nielsen (1968). Panting and blood carbon dioxide in birds. *Am. J. Physiol.* 215: 477–482.

Calder, W.A. (1984). Size, Function and Life History. Cambridge, MA, Harvard University Press, 431 p.

Cameron, J.N. and T.A. Mecklenburg (1973). Aerial gas exchange in the coconut crab, *Birgus latro* with some notes on *Gecarcoidea lalandii*. *Respir. Physiol.* 19: 245–261.

Cameron, J.N. and N. Heisler (1983). Studies of ammonia in the rainbow trout: physico-chemical parameters, acid–base behaviour and respiratory clearance. *J. Exp. Biol.* 105: 107–125.

Campbell, J.W. and L. Goldstein (1972). Nitrogen Metabolism and the Environment. London, Academic Press, 318 p.

Campbell, J.W., J.E. Vorhaben and D.D. Smith (1984). Hepatic ammonia metabolism in a uricotelic treefrog *Phyllomedusa sauvagei*. *Am. J. Physiol.* 246: R805–R810.

Carey, F.G. and J.M. Teal (1969). Regulation of body temperature by the bluefin tuna. *Comp. Biochem. Physiol.* 28: 205–213.

Carey, F.G. (1982). Warm fish. In: A Companion to Animal Physiology, edited by C.R. Taylor, K. Johansen and L. Bolis. Cambridge, Cambridge University Press, 365 p.

Cerretelli, P., R. Sikand and L. Farhi (1966). Readjustments in cardiac output and gas exchange during onset of exercise and recovery. *J. Appl. Physiol.* 21: 1345–1350.

Cerretelli, P. and P.E. Di Prampero (1987). Gas exchange in exercise. In: Handbook of Physiology. Section 3: The Respiratory System. Vol. IV. Gas Exchange, edited by L.E. Farhi and S.M. Tenney. Bethesda, MD, Am. Physiol. Soc. Distributed by The Williams and Wilkins Company, Baltimore, MD, pp. 297–339.

Childress, J.J. and G.N. Somero (1979). Depth-related enzymic activities in muscle, brain and heart of deep-living pelagic marine teleosts. *Mar. Biol.* 52: 273–283.

Childress, J.J., H. Felbeck and G.N. Somero (1987). Symbiosis in the deep sea. *Sci. Am.* 256 (No. 5): 107–112.

Cohen, P.P. (1970). Biochemical differentiation during amphibian metamorphosis. *Science* 168: 533–543.

Coleridge, H.M., J.C.G. Coleridge and R.B. Banzett (1978). Effect of CO_2 on afferent vagal endings in the canine lung. *Respir. Physiol.* 34: 135–151.

Comparative Physiology: Life in Water and on Land (1987), edited by P. Dejours, L. Bolis, C.R. Taylor and E. Weibel. Fidia Research Series, Vol. 9. Padova, Liviana Press, distributed by Springer-Verlag, New York, 556 p.

Conti, E. and A. Toulmond (1986). Ventilation response to ambient hypo- and hypercapnia in the lugworm, *Arenicola marina* (L.). *J. Comp. Physiol. B* 156: 797–802.

Coon, R.L., E.J. Zuperku and J.P. Kampine (1984). Systemic arterial pH servocontrolled ventilator simulation of the respiratory control system. *Respir. Physiol.* 58: 345–350.

Cooper, A.J.L. and F. Plum (1987). Biochemistry and physiology of brain ammonia. *Physiol. Rev.* 67: 440–519.

Coulson, R.A. and T. Hernandez (1983). Alligator metabolism. Studies on chemical reactions *in vivo*. *Comp. Biochem. Physiol.* 74B: 1–182.

Cournand A., D.W. Richards, R.A. Bader, M.E. Bader and A.P. Fishman (1954). The oxygen cost of breathing. *Trans. Assoc. Am. Physicians* 67: 162–173.

Crawford, E.C. (1972a). Brain and body temperatures in a panting lizard. *Science* 177: 431–433.

Crawford, E.C. (1972b). Mechanical aspects of panting in dogs. *J. Appl. Physiol.* 17: 249–251.

Craybrook, D.L. (1983). Nitrogen metabolism. In: The Biologia of Crustacea, edited by D.E. Bliss. Vol. 5, edited by L.H. Mantel. New York, Academic Press, 471 p.

Crompton, A.W., C.R. Taylor and J.A. Jagger (1978). Evolution of homeothermy in mammals. *Nature* 272: 333–336.

Crone, C. (1965). Facilitated transfer of glucose from blood into brain tissue. *J. Physiol. (London)* 181: 103–113.

Crone, C. and N.A. Lassen (1970). Capillary Permeability. The Transfer of Molecules and Ions between Capillary Blood and Tissue. Copenhagen, Munksgaard, 681 p.

Crone, C. (1971). The blood–brain barrier. Facts and questions. In: Ion Homeostasis of the Brain, edited by B.K. Siesjö and S.C. Sørensen. Copenhagen, Munksgaard, pp. 52–66.

Cross, K.W. and D. Stratton (1974). Aural temperature of the newborn infant. *Lancet* No. 11: 1179–1180.

Cserr, H.F. and M. Bundgaard (1984). Blood–brain interfaces in vertebrates: a comparative approach. *Am. J. Physiol.* 246: R277–R288.

Culberson, C. and R.M. Pytkowicz (1970). Oxygen-total carbon dioxide correlation in the Eastern Pacific Ocean. *J. Oceanograph. Soc. Jpn.* 26: 95–100.

Cunningham, D.J.C. (1974). The control system regulating breathing in man. *Q. Rev. Biophys.* 6: 433–483.

Cunningham, D.J.C. (1987). Review lecture. Studies on arterial chemoreceptors in man. *J. Physiol. (London)* 384: 1–26.

Davies, D.G. and J.A. Sexton (1987). Brain ECF pH and central chemical control of ventilation during anoxia in turtles. *Am. J. Physiol.* 252: R848–R852.

DeFur, P.L., P.R.H. Wilkes and B.R. McMahon (1980). Non-equilibrium acid–base status in *C. productus*: role of exoskeletal carbonate buffers. *Respir. Physiol.* 42: 247–261.

Degani, G., N. Silanikove and A. Shkolnik (1984). Adaptation of green toad (*Bufo viridis*) to terrestrial life by urea accumulation. *Comp. Biochem. Physiol.* 77A: 585–587.

Dejours, P., J. Raynaud, C.L. Cuénod and Y. Labrousse (1955). Modifications instantanées de la ventilation au début et à l'arrêt de l'exercice musculaire. Interprétation. *J. Physiol. (Paris)* 47: 155–159.

Dejours, P., A. Teillac, Y. Labrousse and J. Raynaud (1956). Etude du mécanisme de la régulation cardioventilatoire au début de l'exercice musculaire. *Rev. Franç. Etudes Clin. Biol.* 1: 504–517.

Dejours, P. (1959). La régulation de la ventilation au cours de l'exercice musculaire chez l'homme. *J. Physiol. (Paris)* 51: 163–261.

Dejours, P. (1962). Chemoreflexes in breathing. *Physiol. Rev.* 42: 335–358.

Dejours, P. and A. Teillac (1963). Caractères des variations de la ventilation pulmonaire au cours de l'exercice musculaire dynamique chez l'Homme. Etude expérimentale et théorique. *Rev. Franç. Etudes Clin. Biol.* 8: 439–444.

Dejours, P. (1964). Control of respiration in muscular exercise. In: Handbook of Physiology. Section 3. Respiration. Vol. I, edited by W.O. Fenn and H. Rahn. Washington, DC, Am. Physiol. Soc., pp. 631–648.

Dejours, P., R. Puccinelli, J. Armand and M. Dicharry (1965). Concept and measurement of ventilatory sensitivity to carbon dioxide. *J. Appl. Physiol.* 20: 890–897.

Dejours, P. (1966). Respiration. (Translated by L. Farhi). New York, Oxford University Press, 244 p.

Dejours, P., S. Wagner, M. Dejager and M.-J. Vichon (1967). Ventilation et gaz alvéolaire pendant le langage parlé. *J. Physiol. (Paris)* 59: 386.

Dejours, P., J. Armand and G. Verriest (1968). Carbon dioxide dissociation curves of water and gas exchange of water-breathers. *Respir. Physiol.* 5: 23–33.

Dejours, P., W.F. Garey and H. Rahn (1970). Comparison of ventilatory and circulatory flow rates between animals in various physiological conditions. *Respir. Physiol.* 9: 108–117.

Dejours, P. (1973). Problems of control of breathing in fishes. In: Comparative Physiology. Locomotion, Respiration, Transport and Blood, edited by L. Bolis, K. Schmidt-Nielsen and S.H.P. Maddrell. Amsterdam and New York, North-Holland/American Elsevier, pp. 117–133.

Dejours, P. and H. Beekenkamp (1977). Crayfish respiration as a function of water oxygenation. *Respir. Physiol.* 30: 241–251.

Dejours, P., A. Toulmond and J.P. Truchot (1977). The effect of hyperoxia on the breathing of marine fishes. *Comp. Biochem. Physiol.* 58A: 409–411.

Dejours, P. (1979a). La vie dans l'eau et dans l'air. *Pour la Science* No. 20: 87–95.

Dejours, P. (1979b). L'Everest sans oxygène: le problème respiratoire. *J. Physiol. (Paris)* 75: 49A.

Dejours, P. and J. Armand (1980). Hemolymph acid–base balance of the crayfish *Astacus leptodactylus* as a function of the oxygenation and the acid–base balance of the ambient water. *Respir. Physiol.* 41: 1–11.

Dejours, P. (1981). Principles of Comparative Respiratory Physiology, 2nd edition. (First edition 1975). Amsterdam, Elsevier/North-Holland, 265 p.

Dejours, P. (1982). Mount Everest and beyond: breathing air. In: A Companion to Animal Physiology, edited by C.R. Taylor, K. Johansen and L. Bolis. New York, Cambridge University Press, pp. 17–30.

Dejours, P. and J. Armand (1982). Variations of the hemolymph acid–base balance in crayfish as functions of the changes of certain physico-chemical properties of the ambient waters. *C.R. Acad. Sci. Paris* 295, série III: 509–512.

Dejours, P., J. Armand and H. Beekenkamp (1982). The effect of ambient chloride concentration changes on branchial chloride–bicarbonate exchanges and hemolymph acid–base balance of crayfish. *Respir. Physiol.* 48: 375–386.

Dempsey, J.A. and H.V. Forster (1982). Mediation of ventilatory adaptations. *Physiol. Rev.* 62: 262–346.

Dempsey, J.A., E.H. Vidruk and G.S. Mitchell (1985). Pulmonary control systems in exercise: update. *Fed. Proc.* 44: 2260–2270.

Dill, D.B., J.H. Talbott and W.V. Consolazio (1937). Blood as a physicochemical system. XII. Man at high altitude. *J. Biol. Chem.* 118: 649–666.

Donaldson, H.H. (1911). President's address. *J. Neur. Mental Dis.* 38: 257–266.

Doust, J.H. and J.M. Patrick (1981). The limitation of exercise ventilation during speech. *Respir. Physiol.* 46: 137–147.

Dudley, G.A., R.S. Staron, T.F. Murray, F.C. Hagerman and A. Luginbuhl (1983). Muscle fiber composition and blood ammonia levels after intense exercise in humans. *J. Appl. Physiol.* 54: 582–586.

Duellman, W. and L. Trueb (1986). Biology of Amphibians. New York, McGraw-Hill Book Co., 670 p.

Duffin, J. and R.R. Bechbache (1983). The change in ventilation and heart rate at the start of treadmill exercise. *Can. J. Physiol. Pharmacol.* 61: 120–126.

Eclancher, B. and P. Dejours (1975). Contrôle de la respiration chez les Poissons téléostéens: existence de chémorécepteurs physiologiquement analogues aux chémorécepteurs des Vertébrés supérieurs. *C.R. Acad. Sci. Paris* 280: 451–453.

Edney, E.B. and K.A. Nagy (1976). Water balance and excretion. In: Environmental Physiology of Animals, edited by J. Bligh, J.L. Cloudsley-Thompson and A.G. Macdonald. Oxford, Blackwell Scientific Publ., pp. 106–132.

Ekelund, L.G. and A. Holmgren (1964). Circulatory and respiratory adaptation, during long-term, non-steady state exercise, in the sitting position. *Acta Physiol. Scand.* 62: 240–255.

Eldridge, F.L., D.E. Millhorn, J.P. Kiley and T.G. Waldrop (1985). Stimulation by central command of locomotion, respiration and circulation during exercise. *Respir. Physiol.* 59: 313–337.

Elsner, R. and B. Gooden (1983). Diving and Asphyxia. A Comparative Study of Animals and Man. Cambridge, Cambridge University Press, 168 p.

Erasmus, B. de W., B.J. Howell and H. Rahn (1970/71). Ontogeny of acid–base balance in the bullfrog and chicken. *Respir. Physiol.* 11: 46–53.

Faraci, F.M., J.P. Kiley and M.R. Fedde (1984). Chemoreflex drive of ventilation during exercise in ducks. *Pfluegers Arch.* 402: 162–165.

Faraci, F.M., D.L. Kilgore and M.R. Fedde (1985). Blood flow distribution during hypocapnic hypoxia in Pekin ducks and bar-headed geese. *Respir. Physiol.* 61: 21–30.

Farber, J.P. and S.M. Tenney (1971). The pouch gas of the Virginia opossum (*Didelphis virginiana*). *Respir. Physiol.* 11: 335–345.

Farhi, L.E. and H. Rahn (1960). Dynamic of changes in carbon dioxide stores. *Anesthesiology* 21: 604–614.

Farhi, L.E. and D. Linnarsson (1977). Cardiopulmonary readjustments during graded immersion in water at 35 °C. *Respir. Physiol.* 30: 35–50.

Farhi, L.E. (1987). Exposure to stressful environment. In: Comparative Physiology of Environmental Adaptations. Part II: Adaptations to Extreme Environments, edited by P. Dejours. Basel, Karger, p. 224.

Farhi, L.E. (1987). Ventilation–perfusion relationships. In: Handbook of Physiology. Section 3: The Respiratory System. Vol. IV. Gas Exchange, edited by L.E. Farhi and S.M. Tenney. Bethesda, MD, Am. Physiol. Soc. Distributed by The Williams and Wilkins Company, Baltimore, MD, pp. 199–215.

Favier, R., D. Desplanches, J. Frutoso, M. Grandmontagne and R. Flandrois (1983). Ventilatory and circulatory transients during exercise: new arguments for a neurohumoral theory. *J. Appl. Physiol.* 54: 647–653.

Fedde, M.R. and W.D. Kuhlmann (1978). Intrapulmonary carbon dioxide sensitive receptors: amphibians to mammals. In: Respiratory Function in Birds, Adult and Embryonic, edited by J. Piiper. Berlin, Heidelberg, Springer-Verlag, pp. 33–50.

Feder, M.E. and W.W. Burggren (1985). Cutaneous gas exchange in vertebrates: design, patterns, control and implications. *Biol. Rev.* 60: 1–45.

Fenn, W.O., H. Rahn and A.B. Otis (1946). A theoretical study of the composition of the alveolar air at altitude. *Am. J. Physiol.* 146: 637–653.

Fenn, W.O. (1964). Introduction. In: Oxygen in the Animal Organism, edited by F. Dickens and E. Neil. London, Pergamon Press, pp. XI–XVIII.

Fitzgerald, R.S. and G.A. Dehghani (1982). Neural responses of the cat carotid and aortic bodies to hypercapnia and hypoxia. *J. Appl. Physiol.* 52: 596–601.

Flandrois, R., R. Favier and J.M. Pequignot (1977). Role of adrenaline in gas exchanges and respiratory control in the dog at rest and exercise. *Respir. Physiol.* 30: 291–303.

Fordyce, W.E., F.M. Bennett, S.K. Edelman and F.S. Grodins (1982). Evidence in man for a fast neural mechanism during the early phase of exercise hyperpnea. *Respir. Physiol.* 48: 27–43.

Forster, H.V., J.P. Klein, L.H. Hamilton and J.P. Kampine (1982). Regulation of Pa_{CO_2} and ventilation in humans inspiring low levels of CO_2. *J. Appl. Physiol.* 52: 287–294.

Forster, H.V., L.G. Pan, G.E. Bisgard, C. Flynn, S.M. Dorsey and M.S. Britton (1984). Independence of exercise hypocapnia and limb movement frequency in ponies. *J. Appl. Physiol.* 57: 1885–1893.

Freedman, L.S., S. Samuels, I. Fish, S.A. Schwartz, B. Lange, M. Katz and L. Morgano (1980). Sparing of the brain in neonatal undernutrition: amino acid transport and incorporation into brain and muscle. *Science* 207: 902–904.

Funk, G.D. and W.K. Milsom (1987). Changes in ventilation and breathing pattern produced by changing body temperature and inspired CO_2 concentration in turtles. *Respir. Physiol.* 67: 37–51.

Gaillard, S. and A. Malan (1983). Intracellular pH regulation in response to ambient hyperoxia or hypercapnia in the crayfish. *Mol. Physiol.* 4: 231–243.

Garey, W.F. and H. Rahn (1970). Normal arterial gas tensions and pH and the breathing frequency of the electric eel. *Respir. Physiol.* 9: 141–150.

Gatz, R.N., E.C. Crawford and J. Piiper (1974). Respiratory properties of the blood of a lungless and gill-less salamander, *Desmognatus fuscus*. *Respir. Physiol.* 20: 33–41.

Gifford, C.A. (1968). Accumulation of uric acid in the land crab, *Cardisoma guanhumi*. *Am. Zool.* 8: 521–528.

Girard, H., S. Klappstein, I. Bartag and W. Moll (1983). Blood circulation and oxygen transport in the fetal guinea pig. *J. Dev. Physiol.* 5: 181–193.

Glass, M.L. and S.C. Wood (1983). Gas exchange and control of breathing in reptiles. *Physiol. Rev.* 63: 232–260.

Gnaiger, E. and H. Forstner (1983). Polarographic Oxygen Sensors. Berlin, Heidelberg, Springer-Verlag, 370 p.

Goetz, R.H. and E.N. Keen (1957). Some aspects of the cardiovascular system in the giraffe. *Angiology* 8: 542–564.

Goldstein, L. and R.P. Forster (1971). Osmoregulation and urea metabolism in the little skate *Raja erinacea*. *Am J. Physiol.* 220: 742–746.

Gordon, M.S., K. Schmidt-Nielsen and H.M. Kelly (1961). Osmotic regulation in the crab-eating frog (*Rana cancrivora*). *J. Exp. Biol.* 38: 659–678.

Graham, J.B. (1975). Heat exchange in the yellowfin, *Tuna albacares*, and skipjack tuna, *Katsuwonus pelamis*, and the adaptive significance of elevated body temperatures in scombrid fishes. *Fish. Bull.* 73: 219–229.

Graham, J.B., R.H. Rosenblatt and C. Gans (1978). Vertebrate air breathing arose in fresh waters and not in the oceans. *Evolution* 32: 459–463.

Haab, P. (1982). Systématisation des échanges gazeux pulmonaires. *J. Physiol. (Paris)* 78: 108–119.

Haldane, J.S. and J.G. Priestley (1905). The regulation of the lung-ventilation. *J. Physiol. (London)* 32: 225–266.

Hand, S.C. and G.N. Somero (1983). Energy metabolism pathways of hydrothermal vent animals: adaptations to a food-rich and sulfide-rich deep-sea environment. *Biol. Bull.* 165: 167–181.

Hardy, R.N. (1972). Temperature and Animal Life. Studies in Biology No. 35. London, Edward Arnold, 60 p.

Harvey, H.W. (1957). The Chemistry and Fertility of Sea-Waters. Cambridge, Cambridge University Press, 240 p.

Hazelhoff, E.H. (1939). Ueber die Ausnutzung der Sauerstoffs bei verschiedenen Wassertieren. *Z. Vgl. Physiol.* 26: 306–327.

Heath, A.G. (1973). Ventilatory responses of teleost fishes to exercise and thermal stress. *Am. Zool.* 13: 491–503.

Heisler, N. (1982). Intracellular and extracellular acid–base regulation in the tropical freshwater teleost fish *Synbranchus marmoratus* in response to the transition from water breathing to air breathing. *J. Exp. Biol.* 99: 9–28.

Heisler, N., G. Forcht, G.R. Ultsch and J.F. Anderson (1982). Acid–base regulation in response to environmental hypercapnia in two aquatic salamanders, *Siren lacertina* and *Amphiuma means*. *Respir. Physiol.* 49: 141–158.

Heisler, N. (1986). Comparative aspects of acid–base regulation. In: Acid–Base Regulation in Animals, edited by N. Heisler. Amsterdam, Elsevier Science Publishers, pp. 397–450.

Henderson, L.J. (1913). The Fitness of the Environment. New York, The MacMillan Company, 317 p.

Herreid, C.F. (1969a). Water loss of crabs from different habitats. *Comp. Biochem. Physiol.* 28: 829–839.

Herreid, C.F. (1969b). Integument permeability of crabs and adaptation to land. *Comp. Biochem. Physiol.* 29: 423–429.

Herreid, C.F., P.M. O'Mahoney and R.J. Full (1983). Locomotion in land crabs: respiratory and cardiac response of *Gecarcinus lateralis*. *Comp. Biochem. Physiol.* 74A: 117–124.

Heymans, C. and E. Neil (1958). Reflexogenic Areas of the Cardiovascular System. London, Churchill, 271 p.

Hicks, J.W. and S.C. Wood (1985). Temperature regulation in lizards: effects of hypoxia. *Am. J. Physiol.* 248: R595–R600.

Hochachka, P.W. and G.N. Somero (1984). Biochemical Adaptation. Princeton, NJ, Princeton University Press, 537 p.

Hochachka, P.W. and M. Guppy (1987). Metabolic Arrest and the Control of Biological Time. Cambridge, Harvard University Press, 227 p.

Horne, F.R. (1968). Nitrogen excretion in Crustacea. I. The herbivorous land crab *Cardisoma guanhumi* Latreille. *Comp. Biochem. Physiol.* 26: 687–695.

Howell, B.J., F.W. Baumgardner, K. Bondi and H. Rahn (1970). Acid–base balance in cold-blooded vertebrates as a function of body temperature. *Am. J. Physiol.* 218: 600–606.

Howell, B.J., H. Rahn, D. Goodfellow and C. Herreid (1973). Acid–base regulation and temperature in selected invertebrates as a function of temperature. *Am. Zool.* 13: 557–563.

Hughes, G.M., ed. (1975). Respiration of Amphibious Vertebrates. London, Academic Press, 402 p.

Hurtado, A. (1964). Animals in high altitudes: resident man. In: Handbook of Physiology. Adaptation to the Environment. Section 4, edited by D.B. Dill, E.F. Adolph and C.G. Wilber. Washington, DC, Am. Physiol. Soc., pp. 843–860.

Hutchison, V.H., H.G. Dowling and A. Vinegar (1966). Thermoregulation in a brooding female indian python, *Python molurus bivitattus*. *Science* 151: 694–695.

Innes, A.J. and E.W. Taylor (1986). The evolution of air breathing in crustaceans: a functional analysis of branchial, cutaneous and pulmonary gas exchange. *Comp. Biochem. Physiol.* 85A: 621–637.

Ishii, K., K. Honda and K. Ishii (1966). The function of the carotid labyrinth in the toad. *Tohoku J. Exp. Med.* 88: 103–116.

Ishii, K. and T. Oosaki (1969). Fine structure of the chemoreceptor cell in the amphibian carotid labyrinth. *J. Anat.* 104: 263–280.

Ishii, K., K. Ishii and T. Kusakabe (1985). Electrophysiological aspects of reflexogenic area in the chelonian, *Geoclemmys reevesii. Respir. Physiol.* 59: 45–54.

Jackson, D.C. (1971). The effect of temperature on ventilation in the turtle, *Pseudemys scripta elegans. Respir. Physiol.* 12: 131–140.

Jackson, D.C. (1973). Ventilatory response to hypoxia in turtles at various temperatures. *Respir. Physiol.* 18: 178–187.

Jackson, D.C., D.E. Palmer and W.L. Meadow (1974). The effects of temperature and carbon dioxide breathing on ventilation and acid–base status of turtles. *Respir. Physiol.* 20: 131–146.

Jacob, J.S. (1980). Heart rate–ventilatory response of seven terrestrial species of North American snakes. *Herpetologia* 36: 326–335.

Johansen, K. and C. Lenfant (1966). Gas exchange in the cephalopod, *Octopus dofleini. Am. J. Physiol.* 210: 910–918.

Johansen, K. and C. Lenfant (1972). A comparative approach to the adaptability of O_2–Hb affinity. In: Oxygen Affinity of Hemoglobin and Red Cell Acid Base Status, edited by M. Rørth and P. Astrup. Copenhagen, Munksgaard, pp. 750–783.

Johansen, K., A.S. Abe and R.E. Weber (1980). Respiratory properties of whole blood and hemoglobin from the burrowing reptile, *Amphisbaena alba. J. Exp. Zool.* 214: 71–77.

Jones, D.R. and D.J. Randall (1978). The respiratory and circulatory system during exercise. In: Fish Physiology. Vol. VII, edited by W.S. Hoar and D.J. Randall. New York, Academic Press, pp. 425–501.

Jones, D.R. and W.K. Milsom (1979). Functional characteristics of slowly adapting pulmonary stretch receptors in the turtle (*Chrysemys picta*). *J. Physiol. (London)* 291: 37–49.

Jørgensen, C.B. (1975). Comparative physiology of suspension feeding. *Annu. Rev. Physiol.* 37: 57–79.

Jouve-Duhamel, A. and J.P. Truchot (1983). Ventilation in the shore crab *Carcinus maenas* (L.) as a function of ambient oxygen and carbon dioxide: field and laboratory studies. *J. Exp. Mar. Biol. Ecol.* 70: 281–296.

Just, J.J., R.N. Gatz and E.C. Crawford (1973). Changes in respiratory functions during metamorphosis of the bullfrog, *Rana catesbeiana. Respir. Physiol.* 17: 276–282.

Karasov, W.H. and J.H. Diamond (1985). Digestive adaptations for fueling the cost of endothermy. *Science* 228: 202–204.

Keilin, D. (1966). The History of Cell Respiration and Cytochrome. Cambridge, Cambridge University Press, 416 p.

Kellogg, R.H. (1964). Central chemical regulation of respiration. In: Handbook of Physiology. Respiration. Vol. I, edited by W.O. Fenn and H. Rahn. Washington, DC, Am. Physiol. Soc., pp. 507–534.

Kellogg, R.H. (1987). Laws of physics pertaining to gas exchange. In: Handbook of Physiology. Section 3: The Respiratory System. Vol. IV. Gas Exchange, edited by L.E. Farhi and S.M. Tenney. Bethesda, MD, Am. Physiol. Soc. Distributed by The Williams and Wilkins Company, Baltimore, MD, pp. 13–31.

Kelsen, S.G., A. Shustack and W. Hough (1982). The effect of vagal blockade on the variability of ventilation in the awake dog. *Respir. Physiol.* 49: 339–353.

Kiley, J.P. and M.R. Fedde (1983). Cardiopulmonary control during exercise in the duck. *J. Appl. Physiol.* 55: 1574–1581.

Kiley, J.P., F.M. Faraci and M.R. Fedde (1985). Gas exchange during exercise in hypoxic ducks. *Respir. Physiol.* 59: 105–115.

Kinney, J.L., D.T. Matsuura and F.N. White (1977). Cardiorespiratory effect of temperature in the turtle, *Pseudemys floridana*. *Respir. Physiol.* 31: 309–325.

Kinney, J.L. and F.N. White (1977). Oxidative cost of ventilation in a turtle, *Pseudemys floridana*. *Respir. Physiol.* 31: 327–332.

Koch, H.J. and M.J. Hers (1943). Influence des facteurs respiratoires sur les interruptions de la ventilation par le siphon exhalant chez *Anodonta cygnea* L. *Ann. Soc. Zool. Belg.* 74: 32–44.

Kooyman, G.L. (1981). Weddell Seal, Consummate Diver. Cambridge, Cambridge University Press, 135 p.

Kooyman, G.L., M.A. Castellini and R.W. Davis (1981). Physiology of diving in marine mammals. *Ann. Rev. Physiol.* 43: 343–356.

Krieger, J. (1985). Breathing during sleep in normal subjects. *Clin. Chest Med.* 6: 577–594.

Krogh, A. and J. Lindhard (1913). The regulation of respiration and circulation during the initial stages of muscular work. *J. Physiol. (London)* 47: 112–136.

Krogh, A. and I. Leitch (1919). The respiratory function of the blood in fishes. *J. Physiol. (London)* 52: 287–300.

Krogh, A. and J. Lindhard (1919). The changes in respiration at the transition from work to rest. *J. Physiol. (London)* 53: 431–439.

Krogh, A. (1922). The Anatomy and Physiology of Capillaries. New Haven, Yale University Press, 276 p.

Kuffler, S.W., J.G. Nicholls and A.R. Martin (1984). From Neuron to Brain, 2nd edition. Sunderland, MA, Sinauer Associates Inc. Publishers, 651 p.

Lahiri, S., N.S. Cherniack, N.H. Edelman and A.P. Fishman (1971). Regulation of respiration in goat and its adaptation to chronic and life-long hypoxia. *Respir. Physiol.* 12: 388–403.

Lambertsen, C.J., R.H. Kough, D.Y. Cooper, G.L. Emmel, H.H. Loeschcke and C.F. Schmidt (1953). Comparison of relationship of respiratory minute volume to P_{CO_2} and pH of arterial and internal jugular blood in normal man during hyperventilation produced by low concentrations of CO_2 at 1 atmosphere and by O_2 at 3.0 atmospheres. *J. Appl. Physiol.* 5: 803–813.

Langille, B.L. and B. Crisp (1980). Temperature dependence of blood viscosity in frogs and turtles: effect on heat exchange with environment. *Am. J. Physiol.* 239: R248–R253.

Lefrançois, R., H. Gautier and P. Pasquis (1968). Ventilatory oxygen drive in acute and chronic hypoxia. *Respir. Physiol.* 4: 217–228.

Lenfant, C., K. Johansen and G.C. Grigg (1966). Respiratory properties of blood and pattern of gas exchange in the lungfish, *Neoceratodus forsteri* (Krefft). *Respir. Physiol.* 2: 1–21.

Lenfant, C. and K. Johansen (1968). Respiration in the African lungfish *Protopterus aethiopicus*. *J. Exp. Biol.* 49: 437–452.

Levy, M.N. and J.M. Talbot (1983). Carciovascular deconditioning of space flight. *The Physiologist* 26: 297–303.

Lindbom, L. (1983). Microvascular blood flow distribution in skeletal muscle. *Acta Physiol. Scand.* Suppl. 525: 40 p.

Linthicum, D.S. and F.G. Carey (1972). Regulation of brain and eye temperatures by the bluefin tuna. *Comp. Biochem. Physiol.* 43A: 425–433.

Little, C. (1983). The Colonisation of Land. Origins and Adaptations of Terrestrial Animals. Cambridge, Cambridge University Press, 289 p.

Livingstone, D.A. (1963). Data of Geochemistry. Chemical Composition of Rivers and Lakes. Washington, DC, Geological Survey Professional Paper 440-G. United States Government Printing Office, pp. 1–64.

Loewe, R. and H. Brauer de Eggert (1979). Blood gas analysis and acid–base status in the hemolymph of a spider (*Eurypelma californicum*). Influence of temperature. *J. Comp. Physiol.* 134B: 331–338.

Lomholt, J.P. and K. Johansen (1976). Gas exchange in the amphibious fish, *Amphipnous cuchia*. *J. Comp. Physiol.* 107B: 141–157.

Loveridge, J.P. (1970). Observations on nitrogenous excretion and water relations of *Chiromantis xerampelina* (Amphibia, Anura). *Arnoldia* (*Rhodesia*) 5: 1–6.

Loveridge, J.P. and P.C. Withers (1981). Metabolism and water balance of active and cocooned African bullfrogs *Pyxicephalus adspersus*. *Physiol. Zool.* 54: 203–214.

MacMillen, R.E. and A.K. Lee (1969). Water metabolism of Australian hopping mice. *Comp. Biochem. Physiol.* 28: 493–514.

Maddrell, S.H.P. (1971). The mechanism of insect excretory system. *Adv. Insect Physiol.* 8: 199–331.

Malan, A., H. Arens and A. Waechter (1973). Pulmonary respiration and acid–base state in hibernating marmots and hamsters. *Respir. Physiol.* 17: 45–61.

Malan, A. (1977). Blood acid–base state at a variable temperature. A graphical representation. *Respir. Physiol.* 31: 259–275.

Malan, A. (1982). Respiration and acid–base balance state in hibernation. In: Hibernation and Torpor in Mammals and Birds, edited by C.P. Lyman, J.S. Willis, A. Malan and L.C.H. Wang. New York, Academic Press, pp. 237–282.

Malan, A. (1986). pH as a control factor in hibernation. In: Living in the Cold. Physiological and Biochemical Adaptations, edited by H.C. Heller, X.J. Musacchia and L.C.H. Wang. New York, Elsevier, pp. 61–70.

Mangum, C.P. (1977). The analysis of oxygen uptake and transport in different kinds of animals. *J. Exp. Mar. Biol. Ecol.* 27: 125–140.

Marder, J. and I. Gavrieli-Levin (1986). Body and egg temperature regulation in incubating pigeons exposed to heat stress: the role of skin evaporation. *Physiol. Zool.* 59: 532–538.

Massabuau, J.C., B. Eclancher and P. Dejours (1980). Ventilatory reflex response to hyperoxia in the crayfish, *Astacus pallipes*. *J. Comp. Physiol.* 140B: 193–198.

Massabuau, J.C. and B. Burtin (1984). Regulation of oxygen consumption in the crayfish *Astacus leptodactylus* at different levels of oxygenation: role of peripheral O_2 chemoreception. *J. Comp. Physiol.* 155B: 43–49.

Massabuau, J.C., P. Dejours and Y. Sakakibara (1984). Ventilatory CO_2 drive in the crayfish: influence of oxygen consumption level and water oxygenation. *J. Comp. Physiol.* 154B: 65–72.

Massabuau, J.C. and B. Burtin (1985). Ventilatory CO_2 reflex response in hypoxic crayfish *Astacus leptodactylus* acclimated to 20 °C. *J. Comp. Physiol.* 156B: 115–118.

Mazur, P. (1984). Freezing of living cells: mechanisms and implications. *Am. J. Physiol.* 247: C125–C142.

McCaffrey, T.V., R.D. McCook and R.D. Wurster (1975). Effect of head skin temperature on tympanic and oral temperature in man. *J. Appl. Physiol.* 39: 114–118.

McMahon, B.R. and J.L. Wilkens (1977). Periodic respiratory and circulatory performance in the red rock crab *Cancer productus*. *J. Exp. Zool.* 202: 363–374.

McMahon, B.R., D.G. McDonald and C.M. Wood (1979). Ventilation, oxygen uptake and haemolymph oxygen transport, following enforced exhausting activity in the Dungeness crab *Cancer magister*. *J. Exp. Biol.* 80: 271–285.

McNab, B.K. (1978). The evolution of endothermy in the phylogeny of mammals. *Am. Natur.* 112: 1–21.

Meduna, J.L. (1950). Carbon Dioxide Therapy. Springfield, IL, C.C. Thomas, 236 p.

Midtgård, U. (1983). Scaling of the brain and the eye cooling system in birds: a morphometric analysis of the *Rete ophthalmicum*. *J. Exp. Zool.* 225: 197–207.

Milic-Emili, G. and J.M. Petit (1960). Mechanical efficiency of breathing. *J. Appl. Physiol.* 15: 359–362.

Miller, M.J. and S.M. Tenney (1975). Hyperoxic hyperventilation in carotid-deafferented cats. *Respir. Physiol.* 23: 23–30.

Millhorn, D.E., F.L. Eldridge, J.P. Kiley and T.G. Waldrop (1984). Prolonged inhibition of respiration following acute hypoxia in glomectomized cats. *Respir. Physiol.* 57: 331–340.

Milsom, W.K. (1979). The role of pulmonary afferent information and hypercapnia in the control of breathing pattern in chelonia. *Respir. Physiol.* 37: 101–107.

Milsom, W.K. and R.W. Brill (1986). Oxygen sensitive afferent information arising from the first gill arch of yellowfin tuna. *Respir. Physiol.* 66: 193–203.

Mitchell, G.S., T.T. Gleason and A.F. Bennett (1981). Ventilation and acid–base balance during activity in lizards. *Am. J. Physiol.* 240: R29–R37.

Mitchell, H.A. (1964). Investigation of the cave atmosphere of a Mexican bat colony. *J. Mammal.* 45: 568–577.

Morgan, K.R. and G.A. Bartholomew (1982). Homeothermic response to ambient temperature in a scarab beetle. *Science* 216: 1409–1410.

Morris, S. and A.C. Taylor (1985). The respiratory response of the intertidal prawn *Palaemon elegans* (Rathke) to hypoxia and hyperoxia. *Comp. Biochem. Physiol.* 81A: 633–639.

Nagy, K.A. and C.C. Peterson (1987). Water flux scaling. In: Comparative Physiology: Life in Water and on Land. Fidia Research Series, Vol. 9, edited by P. Dejours, L. Bolis, C.R. Taylor and E. Weibel. Padova, Liviana Press, distributed by Springer-Verlag, New York, pp. 131–140.

Nielsen, M. and H. Smith (1951). Studies on the regulation of respiration in acute hypoxia. *Acta Physiol. Scand.* 24: 293–313.

Nomoto, S., W. Rautenberg and M. Iriki (1983). Temperature regulation during exercise in the Japanese quail (*Coturnix coturnix japonica*). *J. Comp. Physiol.* 149: 519–525.

Nonnotte, G. and R. Kirsch (1978). Cutaneous respiration in seven sea-water teleosts. *Respir. Physiol.* 35: 111–118.

Nye, P.C.G. and F.L. Powell (1984). Steady-state discharge and bursting of arterial chemoreceptors in the duck. *Respir. Physiol.* 56: 369–384.

Odum, E.P. (1971). Fundamentals of Ecology, 3rd edition. Philadelphia, W.B. Saunders, 574 p.

Otis, A.B. (1964). The work of breathing. In: Handbook of Respiration. Section 3. Respiration. Vol. I, edited by W.O. Fenn and H. Rahn. Washington, DC, Am. Physiol. Soc., pp. 463–476.

Otis, A.B. (1987). An overview of gas exchange. In: Handbook of Physiology. Section 3: The Respiratory System. Vol. IV. Gas Exchange, edited by L.E. Farhi and S.M. Tenney. Bethesda, MD, Am. Physiol. Soc. Distributed by The Williams and Wilkins Company, Baltimore, MD, pp. 1–11.

Pace, N. and A.H. Smith (1981). Gravity, and metabolic scale effects in mammals. *The Physiologist* 24: S37–S40.

Packard, G.C. (1974). The evolution of air-breathing in paleozoic gnathostome fishes. *Evolution* 28: 320–325.

Pappenheimer, J.R. *et al.* (1950). Standardization of definitions and symbols in respiratory physiology. *Fed. Proc.* 9: 602–605.

Pasquis, P., A. Lacaisse and P. Dejours (1970). Maximal oxygen uptake in four species of small mammals. *Respir. Physiol.* 9: 298–309.

Patterson, J.L., R.H. Goetz, J.T. Doyle, J.V. Warren, O.H. Gauer, D.K. Detweiler, S.I. Said, H. Hoernicke, M. McGregor, E.N. Keen, M.H. Smith, E.L. Hardic, M. Reynolds, W.P. Flatt and D.E. Waldo (1965). Cardiorespiratory dynamics in the ox and giraffe with comparative observations on man and other mammals. *Ann. N.Y. Acad. Sci.* 127: 393–413.

Peeters, L.L.H., R.E. Sheldon, M.D. Jones and E.L. Makowski (1979). Blood flow to fetal organs as a function of arterial oxygen content. *Am. J. Obstet. Gynecol.* 135: 637–646.

Peyraud, C. and A. Serfaty (1964). Le rythme respiratoire de la carpe (*Cyprinus carpio* L.) et ses relations avec le taux de l'oxygène dissous dans le biotope. *Hydrobiologia* 23: 165–178.

Piiper, J., P. Dejours, P. Haab and H. Rahn (1971). Concepts and basic quantities in gas exchange physiology. *Respir. Physiol.* 13: 292–304.

Piiper, J. and P. Scheid (1975). Gas transport efficacy of gills, lungs and skin: Theory and experimental data. *Respir. Physiol.* 23: 209–221.

Piiper, J., M. Meyer, H. Worth and H. Willmer (1977). Respiration and circulation during swimming activity in the dogfish *Scyliorhinus stellaris*. *Respir. Physiol.* 30: 221–239.

Pinshow, B., M.H. Bernstein and Z. Arad (1985). Effects of temperature and P_{CO_2} on O_2 affinity of pigeon blood: implications for brain O_2 supply. *Am. J. Physiol.* 249: R758–R764.

Powers, D.A., H.J. Fyhn, U.E.H. Fyhn, J.P. Martin, R.L. Garlick and S.C. Wood (1979). A comparative study of the oxygen equilibria of blood from 40 genera of Amazonian fishes. *Comp. Biochem. Physiol.* 62A: 67–85.

Powers, L.W. and D.E. Bliss (1983). Terrestrial adaptations. In: Environmental Adaptations. Biology of Crustacea. Vol. 8, edited by F.J. and W.B. Vernberg. New York, Academic Press, 383 p.

Powers, S.K., R.E. Beadle, D. Thompson and J. Lawler (1987). Ventilatory and blood gas dynamics at onset and offset of exercise in the pony. *J. Appl. Physiol.* 62: 141–148.

Prosser, C.L. (1973). Comparative Animal Physiology, 3rd edition. Philadelphia, W.B. Saunders, 966 p.

Prosser, C.L. (1986). Adaptational Biology. Molecules to Organisms. New York, John Wiley and Sons, 784 p.

Rahn, H. and A.B. Otis (1947). Alveolar air during simulated flights to high altitudes. *Am. J. Physiol.* 150: 202–221.

Rahn, H. (1949). A concept of mean alveolar air and the ventilation–blood flow relationships during pulmonary gas exchange. *Am. J. Physiol.* 158: 21–30.

Rahn, H. and W.O. Fenn (1955). A Graphical Analysis of the Respiratory Gas Exchange. The O_2–CO_2 diagram. Washington, DC, Am. Physiol. Soc., 38 p.

Rahn, H. (1966). Aquatic gas exchange: Theory. *Respir. Physiol.* 1: 1–12.

Rahn, H. (1967). Gas transport from the external environment to the cell. In: Development of the Lung. A Ciba Foundation Symposium, edited by A.V.S. de Reuck and R. Porter. London, J. and A. Churchill Ltd., pp. 3–29.

Rahn, H. and C.V. Paganelli (1968). Gas exchange in gas gills of diving insects. *Respir. Physiol.* 5: 145–164.

Rahn, H., K.B. Rahn, B.J. Howell, C. Gans and S.M. Tenney (1971). Air breathing of the garfish (*Lepisosteus osseus*). *Respir. Physiol.* 11: 285–307.

Rahn, H. and W.F. Garey (1973). Arterial CO_2, O_2, pH, and HCO_3^- values of ectotherms living in the Amazon. *Am. J. Physiol.* 225: 735–738.

Rahn, H., R.B. Reeves and B.J. Howell (1975). Hydrogen ion regulation, temperature, and evolution. *Am. Rev. Respir. Dis.* 112: 165–172.

Rahn, H. and R.B. Reeves (1982). Hydrogen ion regulation during hypothermia: from the Amazon to the operating room. In: Applied Physiology in Clinical Respiratory Care, edited by O. Prakash. The Hague, Martinus Nijhoff Publ., pp. 1–15.

Randall, D.J. and J.N. Cameron (1973). Respiratory control of arterial pH as temperature changes in rainbow trout *Salmo gairdneri*. *Am. J. Physiol.* 225: 997–1002.

Raynaud, J., H. Bernal, J.P. Bourdarias, P. David and J. Durand (1973). Oxygen delivery and oxygen return to the lungs at onset of exercise in man. *J. Appl. Physiol.* 35: 259–262.

Reed, D.J. and R.H. Kellogg (1960). Effect of sleep on CO_2 stimulation of breathing in acute and chronic hypoxia. *J. Appl. Physiol.* 15: 1135–1138.

Reeves, R.B. (1972). An imidazole alphastat hypothesis for vertebrate acid–base regulation: tissue carbon dioxide content and body temperature in bullfrogs. *Respir. Physiol.* 14: 219–236.

Reeves, R.B. (1976). Temperature-induced changes in blood acid–base status: pH and P_{CO_2} in a binary buffer. *J. Appl. Physiol.* 40: 752–761.

Reeves, R.B. (1977). The interaction of body temperature and acid–base balance in ectothermic vertebrates. *Annu. Rev. Physiol.* 39: 559–586.

Regnault, M. (1987). Nitrogen excretion in marine and fresh-water crustacea. *Biol. Rev.* 62: 1–24.

Remmers, J.E. and H. Gautier (1972). Neural and mechanical mechanisms of feline purring. *Respir. Physiol.* 16: 351–361.

Revelle, R. (1982). Carbon dioxide and world climate. *Sci. Am.* 247 (No. 8): 33–41.

Riley, J.P. and G. Skirrow (1975). Chemical Oceanography, 2nd edition. Vol. 2. London, Academic Press, 647 p.

Roberts, J.L. (1975). Active branchial and ram ventilation in fishes. *Biol. Bull.* 148: 85–105.

Robin, E.D., H.V. Murdaugh, W. Pyron, E. Weiss and P. Soteres (1963). Adaptations to diving in the harbor seal – Gas exchange and ventilatory response to CO_2. *Am. J. Physiol.* 205: 1175–1177.

Rodeau, J.L. (1984). Effect of temperature on intracellular pH in crayfish neurons and muscle fibers. *Am. J. Physiol.* 246: C45–C49.

Saunders, N.A. and C.E. Sullivan (1984). Sleep and Breathing. New York, Marcel Dekker, 613 p.

Saunders, R.L. (1962). The irrigation of the gill in fishes: II. Efficiency of oxygen uptake in relation to respiratory flow, activity, and concentration of oxygen and carbon dioxide. *Can. J. Zool.* 40: 817–862.

Scheid, P., R.K. Gratz, F.L. Powell and R. Fedde (1978). Ventilation response to CO_2 in birds. II. Contribution by intrapulmonary CO_2 receptors. *Respir. Physiol.* 35: 361–372.

Scheid, P. (1987). Cost of breathing in water- and air-breathers. In: Comparative Physiology: Life on Land and in Water. Fidia Research Series, Vol. 9, edited by P. Dejours, L. Bolis, C.R. Taylor and E. Weibel. Padova, Liviana Press, distributed by Springer-Verlag, New York, pp. 83–92.

Schmidt-Nielsen, K. and P. Lee (1962). Kidney function in the crab-eating frog (*Rana cancrivora*). *J. Exp. Biol.* 39: 167–177.

Schmidt-Nielsen, K. (1964). Desert Animals, Physiological Problems of Heat and Water. Oxford, Clarendon Press, 277 p.

Schmidt-Nielsen, K. (1969). The neglected interface: the biology of water as a liquid–gas system. *Q. Rev. Biophys.* 2: 283–304.

Schmidt-Nielsen, K. (1983). Animal Physiology: Adaptation and Environment, 3rd edition (1st edition 1975, 2nd edition 1979). Cambridge, Cambridge University Press, 619 p.

Schmidt-Nielsen, K. (1984). Scaling: Why is Animal Size so Important? Cambridge, Cambridge University Press, 241 p.

Schoffeniels, E. (1984). Biochimie Comparée. Paris, Masson, 205 p.

Scholander, P.F. (1964). Animals in aquatic environments: diving mammals and birds. In: Handbook of Physiology. Section 4. Adaptation to the Environment, edited by D.B. Dill, E.F. Adolph and C.G. Wilber. Washington, DC, Am. Physiol. Soc., pp. 729–739.

Sébert, P. (1979). Evidence for a central action of CO_2 ventilatory stimulus in Pekin ducks. *J. Physiol. (Paris)* 75: 901–909.

Seymour, R.S. and H.B. Lillywhite (1976). Blood pressure in snakes from different habitats. *Nature* 264: 664–666.

Seymour, R.S. (1987). Physiological correlates of reinvasion of water by reptiles. In: Comparative Physiology: Life on Land and in Water. Fidia Research Series, Vol. 9, edited by P. Dejours, L. Bolis, C.R. Taylor and E. Weibel. Padova, Liviana Press, distributed by Springer-Verlag, New York, pp. 471–481.

Shelton, G., D.R. Jones and W.K. Milsom (1986). Control of breathing in ectothermic vertebrates. In: Handbook of Physiology. Section 3. The Respiratory System. Vol. II. Control of breathing, edited by P.T. Macklem and J. Mead. Bethesda, MD, Am. Physiol. Soc. Distributed by the Williams and Wilkins Company, Baltimore, MD, pp. 857–909.

Shoemaker, V.H., D. Balding, R. Ruibal and L.L. McClanahan (1972). Uricotelism and low evaporative water loss in a South American frog. *Science* 175: 1018–1020.

Shoemaker, V.H. and K.A. Nagy (1977). Osmoregulation in amphibians and reptiles. *Annu. Rev. Physiol.* 39: 449–471.

Shoemaker, V.H., L.L. McClanahan, P.C. Withers, S.S. Hillman and R.C. Drewes (1987). Thermoregulatory response to heat in the waterproof frogs Phyllomedusa and Chiromantis. *Physiol. Zool.* 60: 365–371.

Siesjö, B.K. (1978). Brain Energy Metabolism. Chichester, John Wiley and Sons, 607 p.

Sinha, N.P. and P. Dejours (1980). Ventilation and blood acid–base balance of the crayfish as functions of water oxygenation (40–1500 Torr). *Comp. Biochem. Physiol.* 65A: 427–432.

Somero, G.N. and F.N. White (1985). Enzymatic consequences under alphastat regulation. In: Acid–Base Regulation and Body Temperature, edited by H. Rahn and A. Prakash. Dordrecht, Martinus Nijhoff Publ., pp. 55–80.

Spiess, F.N., K.C. Macdonald, T. Atwater, R. Ballard, A. Carranza, D. Cordoba, C. Cox, V.M. Diaz Garcia, J. Francheteau, J. Guerrero, J. Hawkins, R. Haymon, R. Hessler, T. Juteau, M. Kastner, R. Larson, B. Luyendyk, J.D. Macdougall, S. Miller, W. Normark, J. Orcutt and C. Rangin (1980). East Pacific Rise: hot springs and geophysical experiments. *Science* 207: 1421–1433.

Steffensen, J.F. and J.P. Lomholt (1983). Energetic cost of active branchial ventilation in the sharksucker, *Echeneis naucrates*. *J. Exp. Biol.* 103: 185–192.

Steffensen, J.F. (1985). The transition between branchial pumping and ram ventilation in fishes: energetic consequences and dependence on water oxygen tension. *J. Exp. Biol.* 114: 141–150.

Stevens, E.D. and D.J. Randall (1967). Changes of gas concentrations in blood and water during moderate swimming activity in rainbow trout. *J. Exp. Biol.* 46: 329–337.

Stevens, E.D. (1968). The effect of exercise on the distribution of blood to various organs in rainbow trout. *Comp. Biochem. Physiol.* 25: 615–625.

Stone, H.L., K.J. Dormer, R.D. Foreman, R. Thies and R.W. Blair (1985). Neural regulation of the cardiovascular system during exercise. *Fed. Proc.* 44: 2271–2278.

Studier, E.H., L.R. Beck and R.G. Lindeborg (1967). Tolerance and initial metabolic response to ammonia intoxication in selected bats and rodents. *J. Mammal.* 48: 564–572.

Studier, E.H. and A.A. Fresquez (1969). Carbon dioxide retention: a mechanism of ammonia tolerance in mammals. *Ecology* 50: 492–494.

Sutterlin, A.M. (1969). Effects of exercise on cardiac and ventilation frequency in three species of freshwater teleosts. *Physiol. Zool.* 42: 36–52.

Szlyk, P.C., B.W. McDonald, D.R. Pendergast and J.A. Krasney (1981). Control of ventilation during graded exercise in the dog. *Respir. Physiol.* 46: 345–365.

Taylor, C.R. and C.P. Lyman (1972). Heat storage in running antelopes: independence of brain and body temperatures. *Am. J. Physiol.* 222: 114–117.

Taylor, C.R. and E.R. Weibel (1981). Design of the mammalian respiratory system. *Respir. Physiol.* 44: 1–164.

Taylor, C.R., E.R. Weibel, R.H. Karas and H. Hoppeler (1987). Adaptive variation in the mammalian respiratory system in relation to energetic demand. VIII. Structural and functional design principles determining the limits to oxidative metabolism. *Respir. Physiol.* 69: 117–127.

Taylor, E.W. and M.G. Wheatly (1980). Ventilation, heart rate and respiratory gas exchange in the crayfish *Austropotamobius pallipes* (Lereboullet) submerged in normoxic water and following 3 h exposure in air at 15°C. *J. Comp. Physiol.* 138B: 67–78.

Taylor, E.W. (1982). Control and co-ordination of ventilation and circulation in crustaceans: responses to hypoxia and exercise. *J. Exp. Biol.* 100: 289–319.

Tenney, M. (1979). A synopsis of breathing mechanisms. In: Evolution of Respiratory Processes. A Comparative Approach, edited by S.C. Wood and C. Lenfant. New York, Marcel Dekker, Inc., pp. 51–106.

Tenney, S.M. and D.F. Boggs (1986). Comparative mammalian respiratory control. In: Handbook of Physiology. Section 3. The Respiratory System. Vol. II. Control of Breathing. Part 2, edited by N.S. Cherniack and J.G. Widdicombe. Bethesda, MD, Am. Physiol. Soc. Distributed by The Williams and Wilkins Company, Baltimore, MD, pp. 833–855.

Tercalfs, R.R. and E. Schoffeniels (1962). Adaptation of amphibians to salt water. *Life Sci.* 1: 19–23.

Thomas, S., B. Fiévet, L. Barthélémy and C. Peyraud (1983). Comparison of the effects of exogenous and endogenous hypercapnia on ventilation and oxygen uptake in the rainbow trout (*Salmo gairdneri* R). *J. Comp. Physiol.* 151B: 185–190.

Thornson, T.B., C.M. Cowan and D.E. Watson (1967). Potamotrygon spp.: Elasmobranchs with low urea content. *Science* 158: 375–377.

Toews, D.P. and N. Heisler (1982). The effects of hypercapnia on intracellular and extracellular acid–base status in the toad *Bufo marinus*. *J. Exp. Biol.* 97: 79–86.

Torrance, R.W. (1968). Arterial Chemoreceptors. Oxford, Blackwell Scientific Publ., 402 p.

Toulmond, A. (1975). Blood oxygen transport and metabolism of the confined lugworm *Arenicola marina* (L.). *J. Exp. Biol.* 63: 647–660.

Toulmond, A. (1977). Temperature-induced vatiations of blood acid–base status in the lugworm, *Arenicola marina* (L.). II. *In vivo* study. *Respir. Physiol.* 31: 151–160.

Toulmond, A., P. Dejours and J.P. Truchot (1982). Cutaneous O_2 and CO_2 exchanges in the dogfish, *Scyliorhinus canicula*. *Respir. Physiol.* 48: 169–181.

Toulmond, A. and C. Tchernigovtzeff (1984). Ventilation and respiratory gas exchanges of the lugworm *Arenicola marina* (L.) as functions of ambient P_{O_2} (20–700 Torr). *Respir. Physiol.* 57: 349–363.

Truchot, J.P. (1973). Temperature and acid–base regulation in the shore crab *Carcinus maenas* (L.). *Respir. Physiol.* 17: 11–20.

Truchot, J.P. (1975). Blood acid–base changes during experimental emersion and reimmersion of the intertidal crab *Carcinus maenas* (L.). *Respir. Physiol.* 23: 351–360.

Truchot, J.P. (1975). Changements de l'état acide–base du sang en fonction de l'oxygénation de l'eau chez le crabe, *Carcinus maenas* (L.). *J. Physiol. (Paris)* 70: 583–592.

Truchot, J.P. (1979). Mechanisms of the compensation of blood respiratory acid–base disturbances in the shore crab, *Carcinus maenas* (L.). *J. Exp. Zool.* 210: 407–416.

Truchot, J.P. and A. Duhamel-Jouve (1980). Oxygen and carbon dioxide in the marine intertidal environment: diurnal and tidal changes in rockpools. *Respir. Physiol.* 39: 241–254.

Truchot, J.P. (1981). The effect of water salinity and acid–base state of the blood acid–base balance in the euryhaline crab, *Carcinus maenas* (L.). *Comp. Biochem. Physiol.* 68A: 555–561.

Truchot, J.P. (1984). Water carbonate alkalinity as a determinant of hemolymph acid–base balance in the shore crab, *Carcinus maenas*: a study at two different ambient P_{CO_2} and P_{O_2} levels. *J. Comp. Physiol.* 154B: 601–606.

Truchot, J.P. (1987a). How do the intertidal invertebrates breathe both water and air? In: Comparative Physiology: Life in Water and on Land. Fidia Research Series, Vol. 9, edited by P. Dejours, L. Bolis, C.R. Taylor and E. Weibel. Padova, Liviana Press, distributed by Springer-Verlag, New York, pp. 37–47.

Truchot, J.P. (1987b). Comparative Aspects of Extracellular Acid–Base Balance. Heidelberg, Springer-Verlag, 248 p.

Ultsch, G.R. (1987). The potential role of hypercarbia in the transition from water-breathing to air-breathing in vertebrates. *Evolution* 41: 442–445.

Van Dam, L. (1938). On the utilization of oxygen and regulation of breathing in some aquatic animals. Groningen, Volharding, 143 p.

Vannier, G. (1983). The importance of ecophysiology for both biotic and abiotic studies of the soil. In: New Trends in Soil Biology, edited by P. Lebrun *et al.* Ottignies-Louvain-La-Neuve, Dieu-Brichart, pp. 289–314.

Velasquez, T. (1959). Tolerance to acute anoxia in high altitude natives. *J. Appl. Physiol.* 14: 357–362.

Verdier, B. (1975). Etude de l'atmosphère du sol. Eléments de comparaison et signification écologique de l'atmosphère d'un sol brun calcaire et d'un sol lessivé podzolique. *Rev. Ecol. Biol. Sol* 12: 591–626.

Verna, A., M. Roumy and L.M. Leitner (1975). Loss of chemoreceptive properties of the rabbit carotid body after destruction of the glomus cells. *Brain Res.* 100: 13–23.

Wald, G. (1981). Metamorphosis: an overview. In: Metamorphosis. A Problem in Developmental Biology, edited by L.I. Gilbert and E. Frieden. New York, Plenum Press, pp. 1–39.

Walters, P.J. and L. Greenwald (1977). Physiological adaptations of aquatic newts (*Notophtalmus viridescens*) to a terrestrial environment. *Physiol. Zool.* 50: 88–98.

Wasserman, K., B.J. Whipp and R. Casaburi (1986). Respiratory control during exercise. In: Handbook of Physiology. Section 3. The Respiratory System. Vol. II. Control of breathing. Part 2, edited by N.S. Cherniack and J.G. Widdicombe. Bethesda, MD, Am. Physiol. Soc. Distributed by The Williams and Wilkins Company, Baltimore, MD, pp. 595–619.

Wassersug, R.J., R.D. Paul and M.E. Feder (1981). Cardio-respiratory synchrony in anuran larvae (*Xenopus laevis, Pachymedusa dacnicolor,* and *Rana berlandieri*). *Comp. Biochem. Physiol.* 70A: 329–334.

Weatherley, A.H. (1970). Effects of superabundant oxygen on thermal tolerance of goldfish. *Biol. Bull.* 139: 229–238.

Weibel, E.R., C.R. Taylor, P. Gehr, H. Hoppeler, O. Mathieu and G.M.O. Maloiy (1981). Design of the mammalian respiratory system. IX. Functional and structural limits for oxygen flow. *Respir. Physiol.* 44: 151–164.

Weibel, E.R. (1984). The Pathway for Oxygen. Structure and Function in the Mammalian Respiratory System. Cambridge, MA, Harvard University Press, 425 p.

Weibel, E.R., C.R. Taylor, H. Hoppeler and R.H. Karas (1987). Adaptive variation in the mammalian respiratory system in relation to energetic demand: I. Introduction to problem and to strategy. *Respir. Physiol.* 69: 1–6.

West, J.B. (1983). Climbing Mt. Everest without oxygen: an analysis of maximal exercise during extreme hypoxia. *Respir. Physiol.* 52: 265–279.

West, J.B. and S. Lahiri (1984). High Altitude and Man. Baltimore, MD, Am. Physiol. Soc., The Williams and Wilkins Company, 199 p.

West, J.B. (1986). Man in space. *NIPS* 1: 189–192.

West, N.H. and W.W. Burggren (1982). Gill and lung ventilatory responses to steady-state aquatic hypoxia and hyperoxia in the bullfrog tadpole. *Respir. Physiol.* 47: 165–176.

Wheatly, M.G. and B.R. McMahon (1982). Responses to hypersaline exposure in the euryhaline crayfish *Pacifastacus leniusculus*. I. The interaction between ionic and acid–base regulation. *J. Exp. Biol.* 99: 425–445.

Whipp, B.J. (1981). The control of exercise hyperpnea. In: Regulation of Breathing. Part II, edited by T.F. Hornbein. New York, Marcel Dekker, pp. 1069–1139.

Wichser, J. and H. Kazemi (1974). Ammonia and ventilation: site and mechanism of action. *Respir. Physiol.* 20: 393–406.

Wieser, W. and G. Schweizer (1972). Der Gehalt an Ammoniak und freien Aminosäuren, sowie die Eigenschaften einer Glutaminase bei *Porcellio scaber* (Isopoda). *J. Comp. Physiol.* 81: 73–88.

Wigglesworth, V.B. (1972). The Principles of Insect Physiology, 7th edition. London, Chapman and Hall, 827 p.

Wilkens, J.L. (1982). Respiratory and circulatory coordination in decapod crustaceans. In: Locomotion and Energetics in Arthropods, edited by C.F. Herreid and C.R. Fourtner. New York, Plenum Press Publ. Corp., pp. 277–298.

Wilkes, P.R.H., R.L. Walker, D.C. McDonald and C.M. Wood (1981). Respiratory, ventilatory, acid–base and ionoregulatory physiology of the white sucker *Catostomus commersoni*: the influence of hyperoxia. *J. Exp. Biol.* 91: 239–254.

Womersley, C. (1981). Biochemical and physiological aspects of anhydrobiosis. *Comp. Biochem. Physiol.* 70B: 669–678.

Wood, C.M. and E.B. Jackson (1980). Blood acid–base regulation during environmental hyperoxia in the rainbow trout (*Salmo gairdneri*). *Respir. Physiol.* 42: 351–372.

Wood, S.C. (1971). Effects of metamorphosis on blood respiratory properties and erythrocyte adenosine triphosphate level of the salamander *Dicamptodon ensatus* (Eschscholtz). *Respir. Physiol.* 12: 53–65.

Wood, S.C., R.E. Weber, G.M.O. Maloiy and K. Johansen (1975). Oxygen uptake and blood respiratory properties of the caecilian *Boulengerula taitanus*. *Respir. Physiol.* 24: 355–363.

Wood, S.C. and C. Lenfant (1979). Oxygen transport and oxygen delivery. In: Evolution of Respiratory Processes. A Comparative Approach, edited by S.C. Wood and C. Lenfant. New York, Marcel Dekker, Inc., pp. 193–223.

Wood, S.C., K. Johansen, M.L. Glass and R.W. Hoyt (1981). Acid–base regulation during heating and cooling in the lizard, *Varanus exanthematicus*. *J. Appl. Physiol.* 50: 779–783.

Wood, S.C. (1987). Oxygen transport in aquatic and terrestrial vertebrates. In: Comparative Physiology: Life in Water and on Land. Fidia Research Series, Vol. 9, edited by P. Dejours, L. Bolis, C.R. Taylor and E. Weibel. Padova, Liviana Press, distributed by Springer Verlag, New York, pp. 59–69.

Wood, S.C. and C. Lenfant (1987). Phylogeny of the gas-exchange system: red cell function. In: Handbook of Physiology. Section 3: The Respiratory System. Vol. IV. Gas Exchange, edited by L.E. Farhi and S.M. Tenney. Bethesda, MD, Am. Physiol. Soc. Distributed by The Williams and Wilkins Company, Baltimore, MD, pp. 131–146.

Wright, P.A. and C.M. Wood (1985). An analysis of branchial ammonia excretion in the freshwater rainbow trout: effects of environmental pH change and sodium uptake blockade. *J. Exp. Biol.* 114: 329–353.

Wright, P., T. Heming and D. Randall (1986). Downstream pH changes in water flowing over the gills of rainbow trout. *J. Exp. Biol.* 126: 499–512.

Wygoda, M.L. (1984). Low cutaneous evaporative water loss in arboreal frogs. *Physiol. Zool.* 57: 329–337.

Zapol, W.M., G.C. Liggins, R.C. Schneider, J. Qvist, M.T. Snider, R.K. Creasy and P.W. Hochachka (1979). Regional blood flow during simulated diving in the conscious Weddell seal. *J. Appl. Physiol.* 47: 968–973.

Zapol, W.M. (1987). Diving adaptations of the Weddell seal. *Sci. Am.* 256 (No. 6): 80–85.

Zuntz, N. and J. Geppert (1886). Ueber die Natur der normalen Athemreize und den Ort ihrer Wirkung. *Arch. Gesamte Physiol.* 38: 337–338.

Subject index